中国最新顶尖办公空间

China New Top
Office Space

深圳市创扬文化传播有限公司　编

赵欣　白丹　译

大连理工大学出版社

图书在版编目(CIP)数据

中国最新顶尖办公空间：汉英对照／深圳市创扬文化
传播有限公司编；赵欣，白丹译.—大连：大连理工大学
出版社，2008.6
ISBN 978-7-5611-4115-1

Ⅰ.中… Ⅱ.①深…②赵…③白… Ⅲ.办公室—建筑设
计—中国—汉、英 Ⅳ.TU243

中国版本图书馆CIP数据核字（2008）第056153号

出版发行：大连理工大学出版社
　　　　　　（地址：大连市软件园路80号　邮编：116023）
印　　　刷：利丰雅高印刷（深圳）有限公司
幅面尺寸：230mm×300mm
印　　张：28
出版时间：2008年6月第 1 版
印刷时间：2008年6月第 1 次印刷
策　　划：袁　斌
责任编辑：刘　蓉
责任校对：倪春子
特约编辑：张长江
封面设计：温广强

ISBN 978-7-5611-4115-1
定　　价：228.00元

电　话：0411-84708842
传　真：0411-84701466
邮　购：0411-84703636
E-mail: dutp@dutp.cn
URL: http://www.dutp.cn

□ **Contents** 目录

GARCIA

加西亚

地　　点：广州
面　　积：900平方米
设 计 师：张星
设计单位：香港东仓设计策划顾问有限公司

Since the original structure of the construction was too narrow and too small, with the analysis of the structure designer, the crossbeam is demolished and more poles are added, thus it forms a cylinder space of the existing great hall. The ample dimensions enable the space to have an intense visual extension. The space is fully manifested in the key tone of dark gray. The indirect lighting of floor lamps nicely produces a balanced ejection of the light on the spatial wall, and explicitly outlines a three-dimensional effect of the wall. Moreover, to better avoid the influence of the dazzling light in the space and to better balance the soft lights, the designer pays much attention on not using materials of high reflection.

由于原建筑结构过于狭小，经结构设计师配合分析，拆除室内横梁并加建柱位，形成现有大堂圆筒形空间，足够的尺度使得空间具有强烈的视觉张力。配合深灰色基调使用更使得空间感得到充分体现。地灯间接照明手法很好地使光效均衡弹射于空间墙体上，并更明确地勾勒出墙体立体效果。另外，为了更好地避免空间中炫光的影响，在材料的选择上尽量避免使用高反光材质，同时也有利于漫光线的均衡。

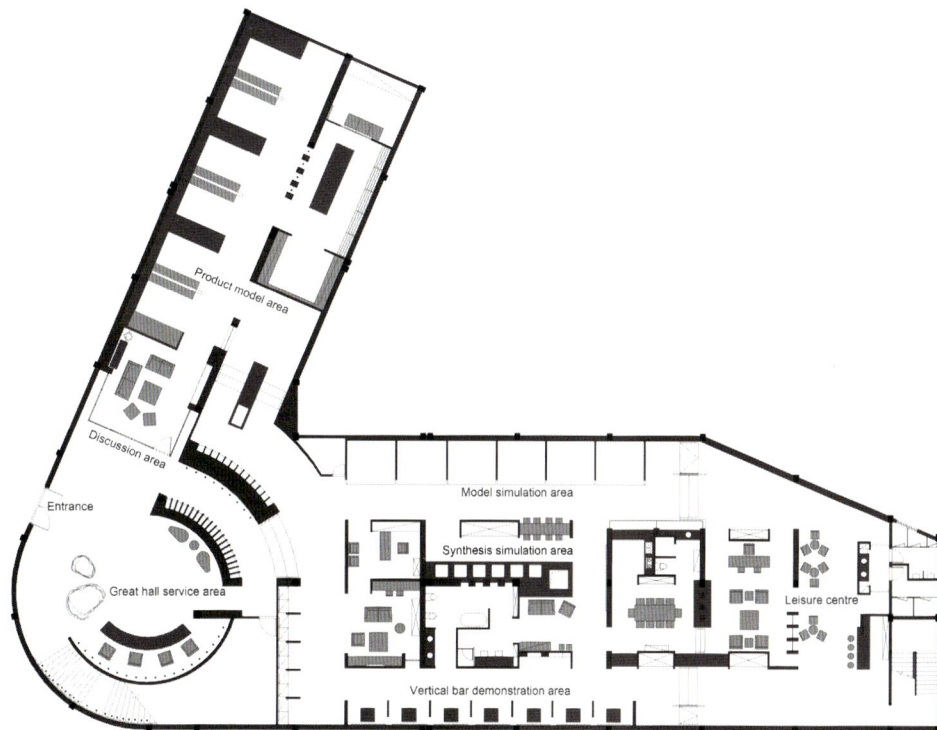

Product model area

Discussion area

Entrance

Great hall service area

Model simulation area

Synthesis simulation area

Leisure centre

Vertical bar demonstration area

LOLA

楼 兰

地　　点：广州
面　　积：1800平方米
设 计 师：张星
设计单位：香港东仓设计策划顾问有限公司

The Designer trys to re-act oriental emotions in the project through both contemporary technologies and general tastes.

There is a desert land in the northwest of China, LOLA, and like Pompeii, it is full of mystery and advanced civilization in history but nobody knows what it should be like.

Now we have not any evidence to rebuild the old LOLA, but some legends and indirect informations can tell us keywords to describe it: wind and dust, mirage, religions, and houses made of stones... All impressions are the only basises for us to realize a space. However, today's aesthetic values, the mainstream thinking and technologies are far different from those of the old years, so designers have to try to rebuild the LOLA in their mind with the characteristic wisdom of the oriental and modern technology. Analyzing the representative elements of China's civilization, we can abstract black and yellow from official uniforms in ancient China for the color, sand pools from landforms for the shape, square and round from laws for the scale and calmness from traditional Chinese paintings for the spirit. Finally, designers are able to sketch a new space with these found elements, in which the spirits are needed more greatly than the essence of the vision.

设计师试图在这个项目中用当代的技术手段和普遍审美重新演绎东方的情绪。

中国的西北有一片被沙漠化的土地，历史中它像庞贝一样充满了神秘的气息和高度的文明。没有人知道它应该是什么样子，它叫做楼兰。

如今我们没有任何的证据来重现当年的楼兰景象，只有一些传说和间接信息告诉我们它的关键词：风沙，海市蜃楼，宗教，石屋……

所有的印象是需要实现一个空间的惟一依据，然而今天的审美、主流思维和技术已经远离了那个年代，于是设计师努力地使用东方人特有的智慧和当代的工艺来再现他们心中的楼兰。

解构中国文明中的代表性元素，色彩从官袍中提炼出黑和黄，形态从地貌中提炼出沙池，尺度从法则中提炼出方圆，精神从国画中提炼出〝平静〞。最终设计师用寻找到的元素具象出一个新空间，精神需求大于视觉本质的空间。

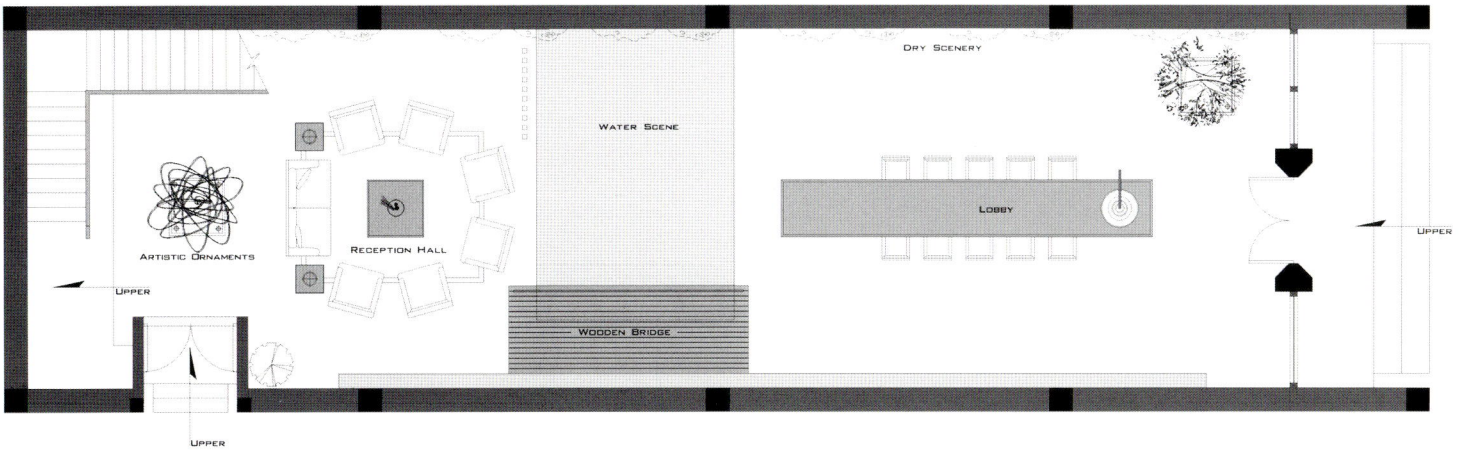

ARTISTIC ORNAMENTS

RECEPTION HALL

WATER SCENE

DRY SCENERY

LOBBY

UPPER

UPPER

WODDEN BRIDGE

UPPER

UPPER

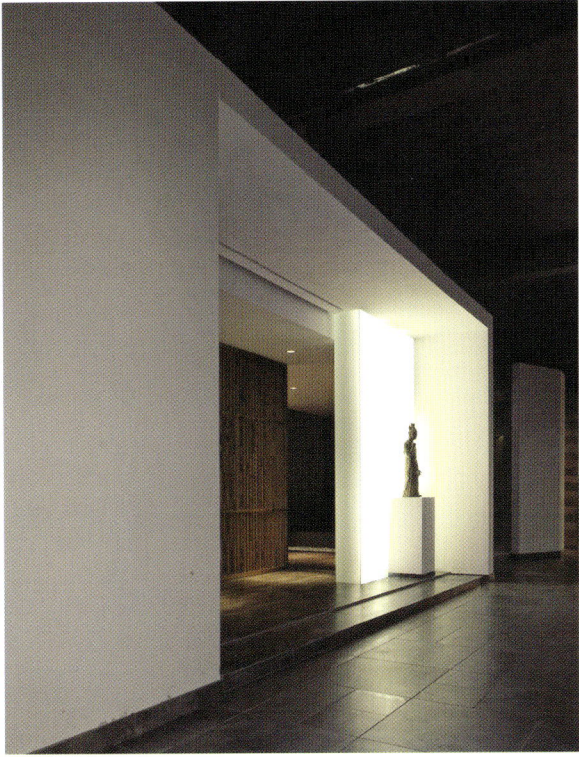

Natural wood-like family

原生态木纹世家

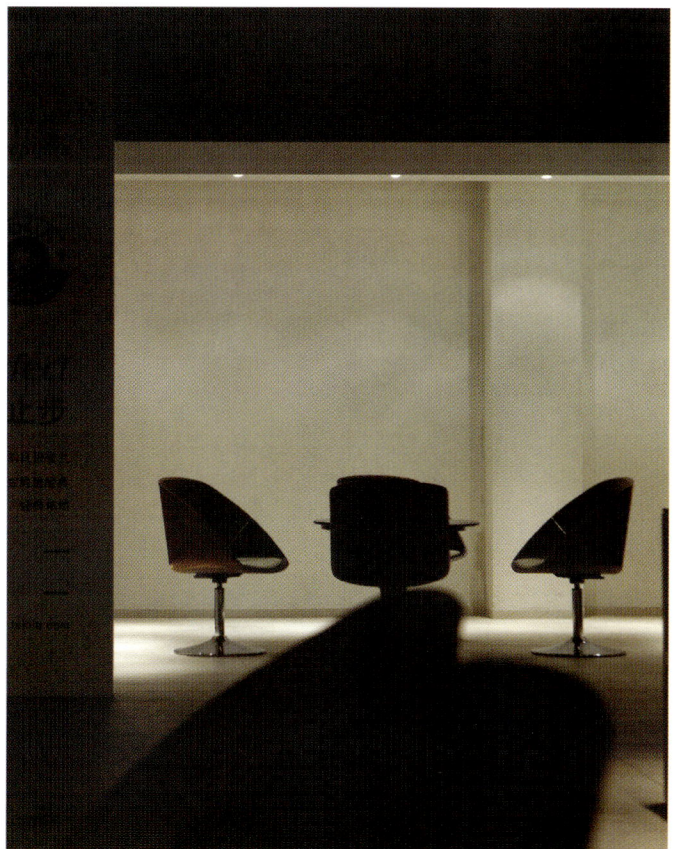

Kenneth Ko Designs Ltd. Shenzhen Office

高文安设计公司深圳办公室

地　　点：深圳
面　　积：1200平方米
设 计 师：高文安
设计单位：高文安设计公司

As an industrial town which is of military importance in Shenzhen, some old factories in the east of Huaqiaocheng Industrial Area is seeking a new way to develop as time goes. With the respect of the old architectures, the design idea of "City Loft" becomes very popular among some designers. Through more than one year's hard work, Mr. Kenneth Ko rented one of the old factory buildings whose height is nearly 8 meters, with an area of nearly 800 square meters. He reconstructed it into an office with the coverage area of nearly 1200 square meters. By means of this special way, Mr. Ko saved this precious industrial heritage and made it become one part of the city's historical development.

The old factory building after renovation and reconstruction becomes a new space which is a multi-functioned one for office, gym, meeting and academic communication. The first floor is reconstructed as swimming pool, gym and staff's dinning room, material room and meeting room, etc. The second floor is reconstructed as the designers' office which has a capacity of about 80 people. The third floor is reconstructed as Mr. Ko's private space. The trilateral steps not only serve as the way to the second floor, but also as the bench for the meeting room and the exhibit stand of artistic products.

On the basis of "reconstructing as the old style", the designer adopts industrial materials which are in harmony with the original architecture, and adopts architectural materials like old red bricks, Chengdu sleepers and Shanghai Stone Gates, which make the new architecture not only keep the original surface, but also increase the inner space's artistic expression, interest, and sense of the new age.

作为深圳工业重镇的华侨城东部工业区，由于时代发展的原因，老工厂已开始谋求新的生存与发展。基于对这些老旧建筑的尊重与爱护，"城市LOFT"为一批设计师所青睐。高文安先生经过一年多的努力，租下了其中一栋高近8米、面积达800平方米的厂房，将其改建成约为1200平方米的办公室。利用这一特殊的形式，使这份珍贵的工业遗产得以保存，并成为城市历史发展的一部分。

设计改造后的老厂房已变成集办公、健身、聚会和学术交流为一体的新空间，一层开辟为游泳池、健身房、员工餐厅、材料室以及会议室等用途，二层为能容纳80人左右的设计师办公室，三层为高先生本人的私人空间。会议室的三面台阶，既可以通往二层，又可以作为会议室的坐台，还可以作为艺术摆件的展架。

设计本着"整旧如旧"的原则，采用与原建筑相协调的一些工业材料，并采购了一批旧红砖、成都枕木和上海的石库门等建筑材料，使改建后的建筑不仅保持了工厂原有的外貌，同时使内部场所增添了艺术性、趣味性和新的时代感。

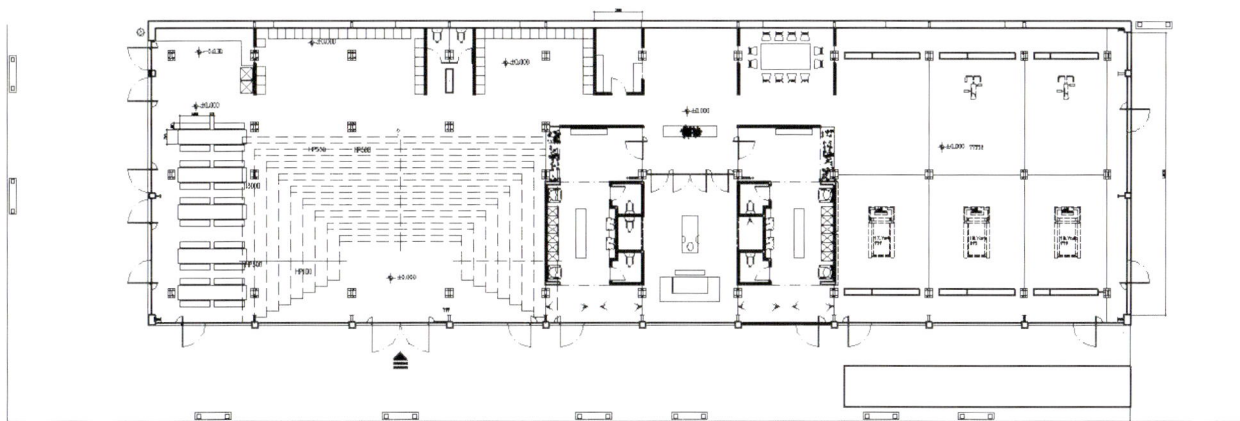

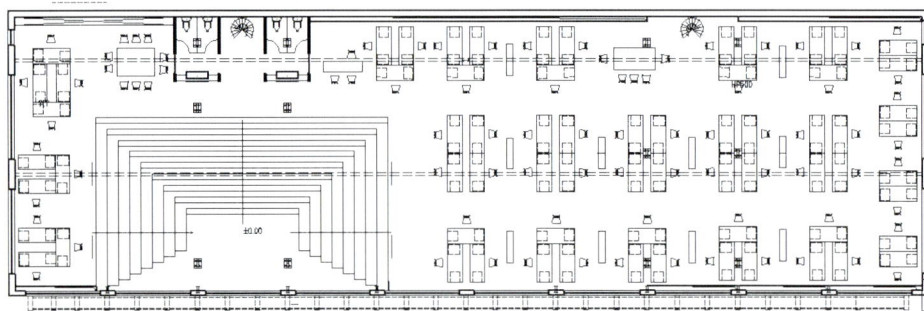

Niefeng Workroom

聂风工作室

地　　点：深圳
面　　积：720平方米
设 计 师：聂剑平
设计单位：深圳市世纪雅典居装饰工程有限公司

The designer makes this small and narrow place which can be seen through easily into an interesting and spacious place, and makes the whole space as a polygon which will not be visually bald. A trapeziform drawing room is left in the center to keep the open sense of the LOFT as soon as you come through the door. The space on the first floor is fluent and has multi-visual angles. On the second floor, the three walls of the trapeziform inner empty place go around and form a working place. The third floor is hidden in a place which can not be seen very easily. There is no sense of pressure about the space. At the same time, the use of space is enlarged and maximized.

What the design shows us is an entia of antinomies which both have very modern space perception and very dense indigenous culture. From the door to the open trapeziform space, there are landscapes of Tibet, Shanxi door building of the atrium, ancient boat of Lijiang, the transparent glass of the inside building, the stairs without armrest, the bluestone floor and plain concrete wall that all show a kind of multi-cultural communication scene. The nine red and yellow pillars support the plane steel balcony, which also shows the imbalanced cultural conflicts. They are a harmony of antinomies.

　　将一个原本不太大而且一眼望穿的局促空间做得既有趣又空旷，将整个空间规划成多边形，视觉上不单调，同时中间尽最大可能留了一个梯形"客厅"，以保证LOFT的空旷之感在进门伊始尽现眼前。一楼的空间是流畅、多视角的。二楼沿梯形中空的三边墙围合布置工作空间。三楼隐藏在一个不起眼的位置，没有空间的压迫感，又最大化利用了空间。

　　在设计中展示出来的是一种矛盾的统一体，既有非常现代的空间感，又有很浓重的本土文化。从大门开始到梯形中空开阔的空间，墙上的西藏风情照片，中庭的山西门楼，丽江的古船，室内建筑本身通透的玻璃，没扶手的楼梯，青石砖的地面，素混凝土的墙，都展示着一种多元文化交流的场景，而外墙的九根红、黄色柱子支撑着水平钢板的阳台，无不演绎着不均衡的文化冲突，是一种矛盾着的和谐统一。

Juzhong Décor Zhongshan Branch

居众装饰中山分部

地　　点：中山
面　　积：300平方米
设 计 师：甘思南
设计单位：居众装饰

To an energetic and creative design company, common design of the space will make it vapid and lack of challenge. Based on this, the designer uses dummy method to deal with this space. The gray mirror plays an important role in this space. It has double increased the space of the original one which is not very large. Under the reflection of the gray mirror, the clipper-built sculpt ceiling makes the space more transparent. Local naked girder structure increases the cramped height of the space, and the gray black ceiling has played a role of harmony and makes the space atmosphere steady.

Gray, white and black should not be parts of colors. The designer uses these three colors as the space color deliberately to provide workers a quiet working environment. The red sofa, curtain and background wall mobilize the colorless space. The classical combination of red and gray white black puts some impacting elements to workers.

On the elevation face, the different camber discontinuous wall, the geometric evolvement between face to face and the partitions in the walkway can be showed adequately which you will never forget as soon as you see it. The use of different materials between two surfaces, the combination of color, virtuality and truth of the space all enrich the space by light from different directions and at the same time enable the light to move in it. So, such an intelligent place which is simple but not lack of interests and connotation has been shown in front of us.

对于充满活力和创造力的设计公司来说，常规的空间布局自然会显得了无生趣以及缺乏挑战性。基于这一前提，设计师采用了虚拟的手法来塑造这一空间，灰镜在这一空间里担当起至关重要的角色，为原本面积不多的空间实现了空间价值的"平方"。

流线型的造型吊顶在灰镜的反射下使空间更具穿透力，部分暴露式梁架增加了原本显得局促的层高，灰黑色顶也起到调和以及稳定空间氛围的作用。

灰、白、黑，本不在色彩之中，设计师特意以黑白灰作为空间色彩，为的是给工作者提供一个能够静心工作的环境。红色沙发、红色纱帘、红色形象背景墙，将无彩色调子的空间调动起来，红色与灰白黑这一经典组合，给工作者注入了冲击性元素。

立面上不同弧度间断斜墙，面与面的几何演变，在走道空间的几个隔断都能得到充分的解说，让人过目难忘。面与面不同材质的运用，色彩的搭配以及空间的虚与实，都在不同角度的灯光的映衬下，丰富了空间，也让光线在其间游走。于是，这样一处简洁大气又不乏趣味和内涵的灵性空间，就在我们面前展现出来。

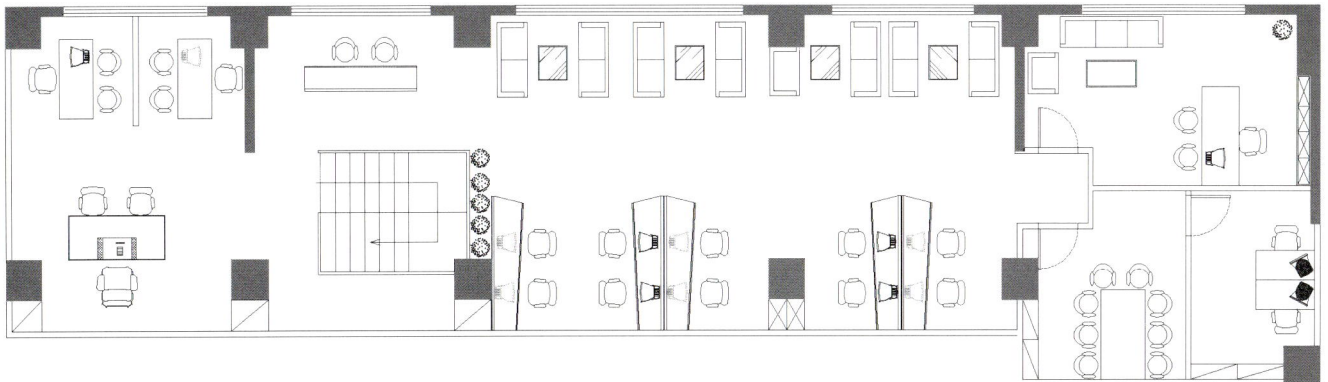

Show Day Office

展迪办公室

地　　点：南京
面　　积：500平方米
设 计 师：李浩澜
设计单位：展迪.浩澜工作室

The space is designed as an office of the designer and his companions. So it is a place which belongs to the designers. The task of this office is not to attract people by a special kind of focus but to create an atmosphere. That is to keep your heart equable which can help you finish your design better. So, the combination of modesty and simplicity, massiveness and quietness is the effect the designer wants in this design.

Emphasis is placed on the organization and the structure of the space, from the outside to the inside, from dynamic to static, from fully transparent to half enclosed to fully close, from the furniture exhibited to natural light, lamplight and liquid body. On the usage of design elements, the material is completely the simplest and the most essential one. All the walls and ground are covered by concrete only. The meeting room on the first floor is like a box fixed up by wooden materials in this open space, so it has the tincture of mellowness. And this is the shining point of the whole space. The color is a combination of black, white and gray. The treatment of the light is very important as well. These three low-pitched colors and the proper light are what the atmosphere should be.

Though the treatment of the space is simple, some details give us a kind of over-sized feeling. A large area of concrete floor, extremely lathy tea table and sofa, and high suspended ceiling and girder and so on are all over-sized experiences. Such kind of experience exists in a static situation. Working in such kind of office, you can get down to thinking and can start a communication between heart and nature.

这个空间是作为设计师和同伴们的办公室而产生，是一个属于设计师的地方，所以它所被赋予的任务，不是靠某种独特的"焦点"来吸引外界目光，而是营造一种氛围，让心灵沉淀下来，更好地完成设计工作。所以，质朴与简约、厚重与宁静的并存才是设计师最想要的结果。

在空间的组织结构上，强调了由外及内，由动及静，由通透到半围合到封闭，由陈列的家具到自然光线、灯光和水体。在设计元素的运用上，却完全是最质朴最本质的材料。所有的墙体和地面都只用了水泥做覆盖。作为一楼大厅的会议室，像是以木质的材料在整个敞开的空间中包裹出的一个盒子，具有柔美的气息。这也正是整个空间的亮点所在。空间的色彩以黑、白、灰相结合。光线的处理也有着极其重要的作用。低调的三种色彩，恰到好处的灯光，这才是氛围的所在。

虽然空间的处理方法简单，但是很多细节却带给人一种超尺度的感受。大面积的水泥地面，极细长的茶几和沙发，高悬的天花和门梁……这种尺度体验，是在一种静态情境下存在的。在这样的办公室里工作，可以静心思考，随时可以开始心灵与自然的对话。

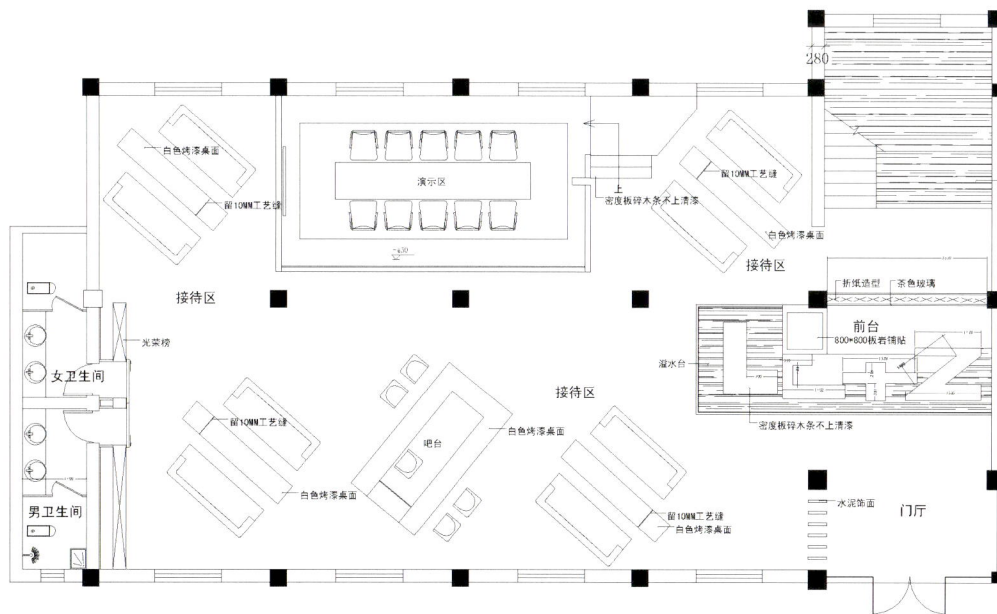

Guoguang Yiye Dianfang Zhiye Building Office

国广一叶点房置业写字楼办公室

地　　点：福州
面　　积：1500平方米
设 计 师：叶斌 叶猛
设计单位：福建国广一叶装饰机构

A Special Space Game Between Culture and Business

This project is bold in the employment of materials, unique in design and grand in style. It places emphasis on humanity, individualism and artistry, displaying a perfect combination of business space and cultural environment. The layout of the space is briefly partitioned, its generatrix design is smooth, and its local places are of unique originality and variation in patterns. The space accords with modern elites' business needs.

Red and black, keen Chinese cultural elements, present a classical and grand style. Emphasis is placed on the design of psychological environment, intelligent environment and cultural environment, and stress is put on an omnidirectional experience of substances, spirits, leisure and recreation. The space stinks with the materials of steel, stone and wood in an innovative intrepidity. The generatrix design between the lobby and office area is smooth, embodying the openness, fluxion and diversity of the business space. A vast and grand business space scales are created in the architecture with the employment of the black ceiling, strong lighting, tracery windows, the mirror glass and the staggered-floor living room. A special space is created in the game of office environment with culture and business, and a business service background of graveness and nobleness can be felt at any time.

文化与商业的另类空间博奕

　　本案用材大胆、设计独特、风格大气，注重人性化、人文化、艺术化，体现了商业空间与文化环境的完美结合。空间格局分割简洁，动线设计流畅，局部匠心独具，富有变化，极为符合现代精英人士的商业需求。

　　强烈的中国文化元素，红与黑体现着古典大气的风格。注重心理环境、智能环境、文化环境的设计，提倡物质与精神、休闲与娱乐的全方位体验。装饰上大胆采用钢、石、木等材料，空间张扬。大堂与办公区域之间动线流畅，体现着商业空间的开放、流动、多元性；黑色天花板、高度灯光、镂空格窗、镜面玻璃、错层会客区，在建筑中创造更为广阔大气的商业空间尺度；时时感受着庄重、高贵的商业服务背景，是办公环境在文化与商业的博奕过程中产生的另类空间。

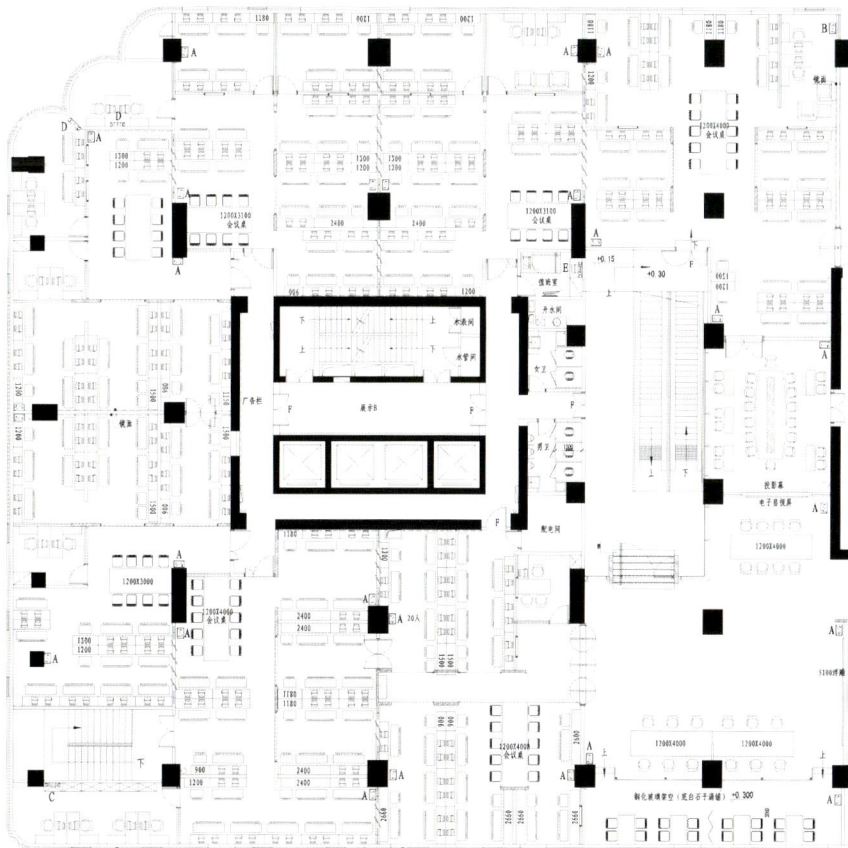

Guangzhou Linguaphone Language Training Center

广州灵格风语言教育中心

地　　点：广州
面　　积：1500平方米
设 计 师：黄志达
设计单位：黄志达设计顾问（香港）有限公司

An Inexhaustible Illusion of The Ocean and The Sky

Linguaphone is a famous language training institution all over the world. Its base in Guangzhou attracts people as well. The space design of this training center uses fashionable, simple, elegant and free style to create an international activity-oriented learning center. Its main function is learning and entertainment. So the designer did not make it a regular and traditional one as the normal studying place. On the contrast, he adapts some easy expressing methods. The inspiration of the design comes from the ocean and the sky. But it does not mean to move the ocean and the sky to it, but to pick up the representative colors instead. Blue is the main color together with many other vivid colors. The colors of red, yellow, orange and green stand for the pure heart of children. The round carpet pattern, ceiling droplight, wave pattern wall, coral shaped chairs as well as some local mucus-made seaweed style and green light provide an imaging space of the sea and the outer space. The infinitude of the ocean and outer space stands for the infinitude of the learning space.

The designer pays more attention to the use of the inside room light effects. It uses baffling LED light, fabricating lamp trough and main wall decoration to increase the interest of learning and reduce the vapidity appeared during learning. The high effective learning can be achieved by playing through learning and learning through playing.

海洋与天空的无穷梦幻

　　灵格风，一个风靡全球的著名语言培训机构，其中国广州基地同样引人注目。其广州培训中心的空间设计秉承了时尚、简约、优雅、自由自在的整体风格，目的是打造一个国际活动式的学习中心，主要功能是学习和娱乐，所以设计师并没有按照正规学习场所那样布置得规整而传统，相反采用一些比较轻松的表现手法。

　　设计灵感取自于海洋和太空，但并不是将海洋和太空的实物生搬硬套上去，而是取其中代表性的颜色。蓝色为主色调，并配以其他鲜艳的颜色。红、黄、橙、绿等代表儿童天真、纯洁的心灵，圆形的地毯图案、天花的吊灯、海浪纹的墙身以及像珊瑚一样的小椅子配以局部阿加力胶制造的海草款式以及草绿色灯光效果，为小朋友们提供游刃于海洋和太空中的想象空间，海洋和太空的无尽无尽同样也象征着学习的空间是无止境的。

　　室内注重灯光效果的运用，采用变幻的LED灯、制造灯槽以及各主墙修饰，同样是为了加强学习的趣味性，达到减少小学生上课产生的乏味性，在学习中玩耍，在玩耍中学习，高效率学习的目的。

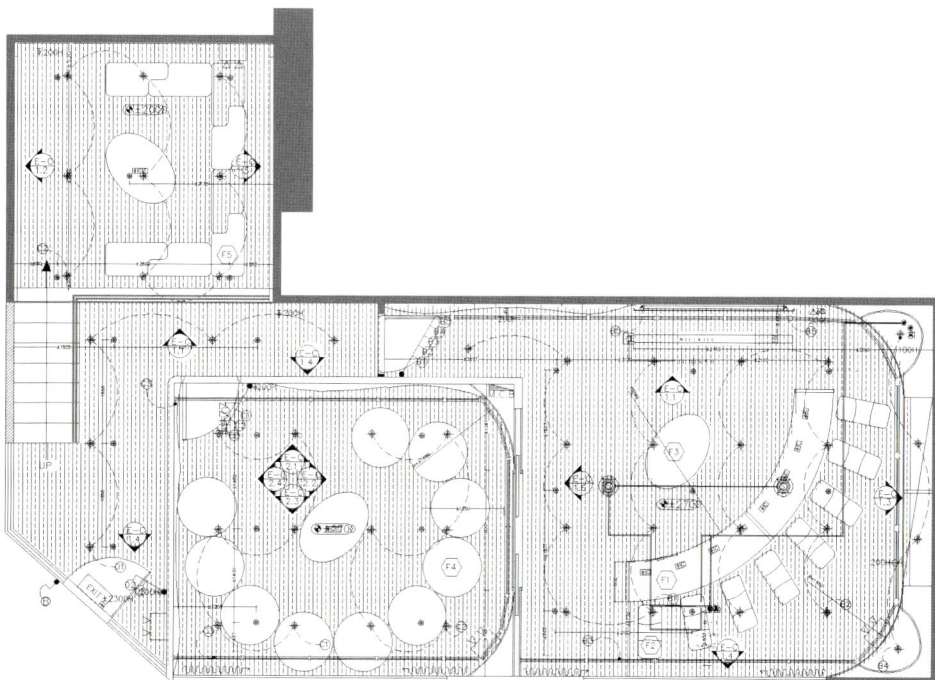

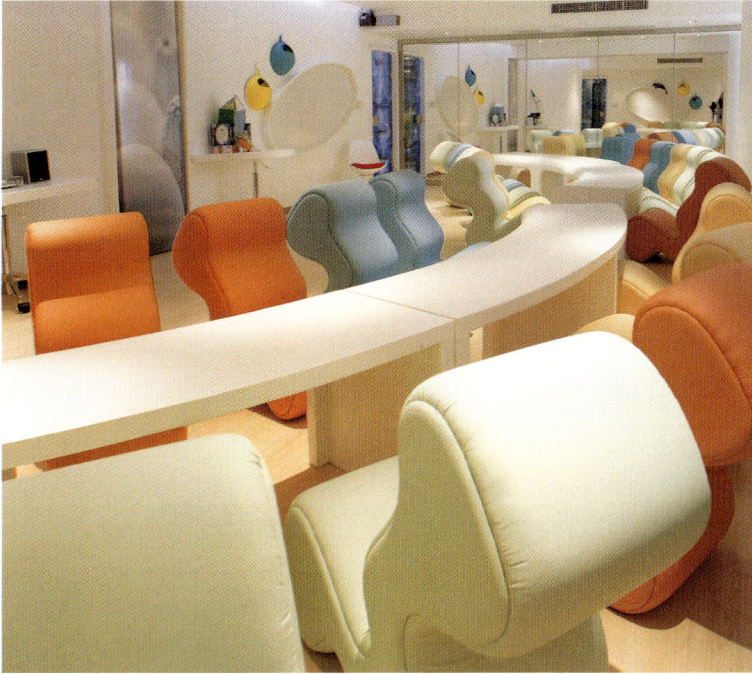

Artapower New Office in Jiangmen

万代华艺美文化用品公司（江门）

地　　点：江门
面　　积：36000平方英尺
设 计 师：蔡明治
设计单位：蔡明治设计有限公司

Artapower deals with the design and manufacture of products. Different from ordinary manufacturers, it provides product design for customers. Its workshop and office building located in Jiangmen is a three-storied office building with a floor area of 36,000 square feet. Its first floor is for the reception area, customers waiting area and a small meeting room. The second floor is for the staff area and department managers' office area along with both a small meeting room and a large one. The top floor is the products exhibition room for customers, the meeting room and the collective thinking room and reading room exclusively for products designers. In the design, the whole picture is enriched by colors, calligraphic elements are employed as one of the marking methods, which reinforces originality and manifests uniqueness as well.

　　万代华艺美文化用品公司是产品设计及生产公司。与一般生产商不同的是，其为顾客提供产品设计的服务。其设于江门的厂房及办公楼。办公楼楼高3层，面积约36000平方英尺。首层为接待区、候客区和小型会议室。二层为职员及各部门经理的办公区，设有小型及大型会议室。顶层主要是招待客人参观的产品展览室、会议室、给产品设计师专用的集思房及图书室。设计方面利用颜色丰富了整个画面，运用书法的元素为其中一种标示方式，加强了创意，彰显了个性。

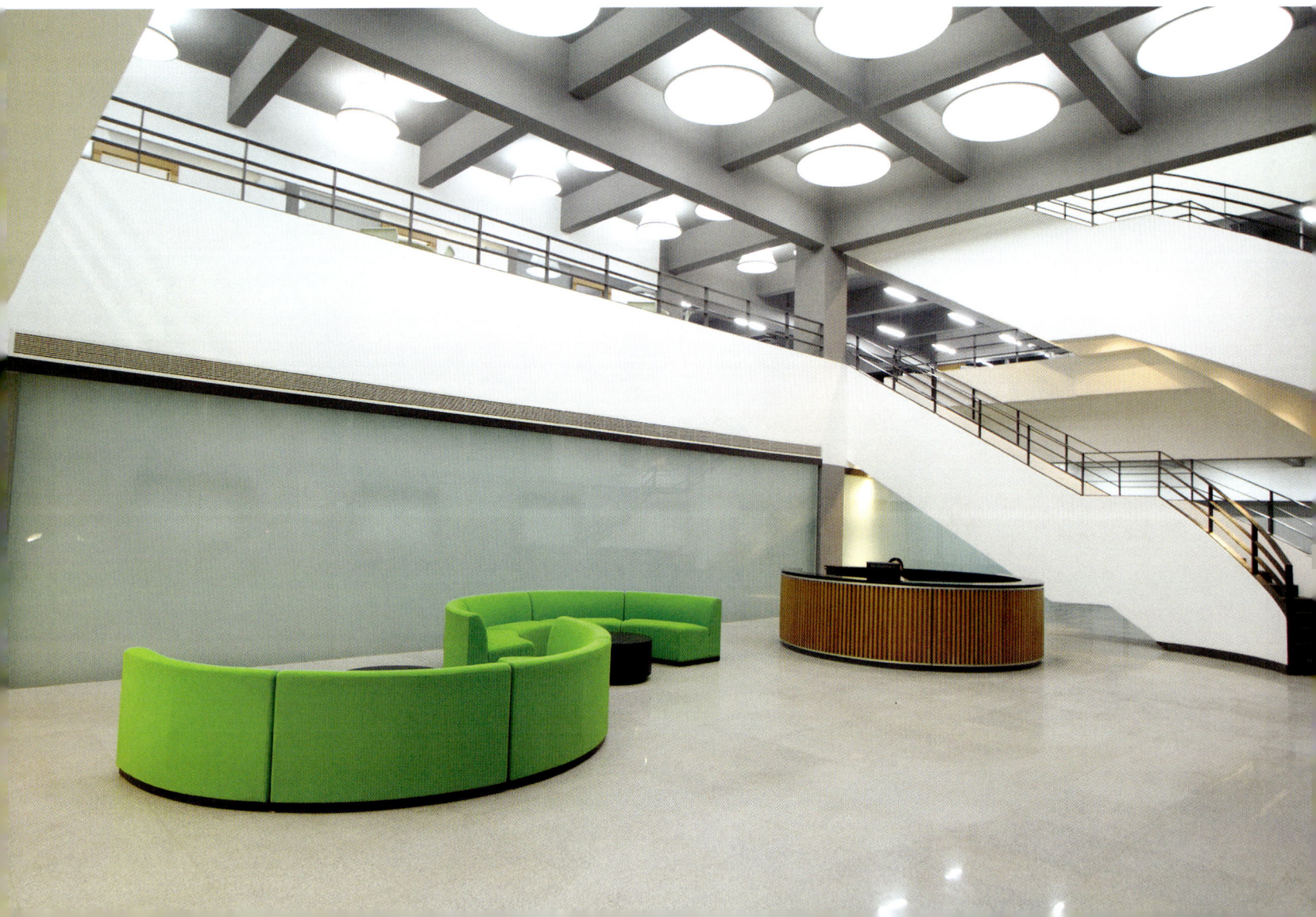

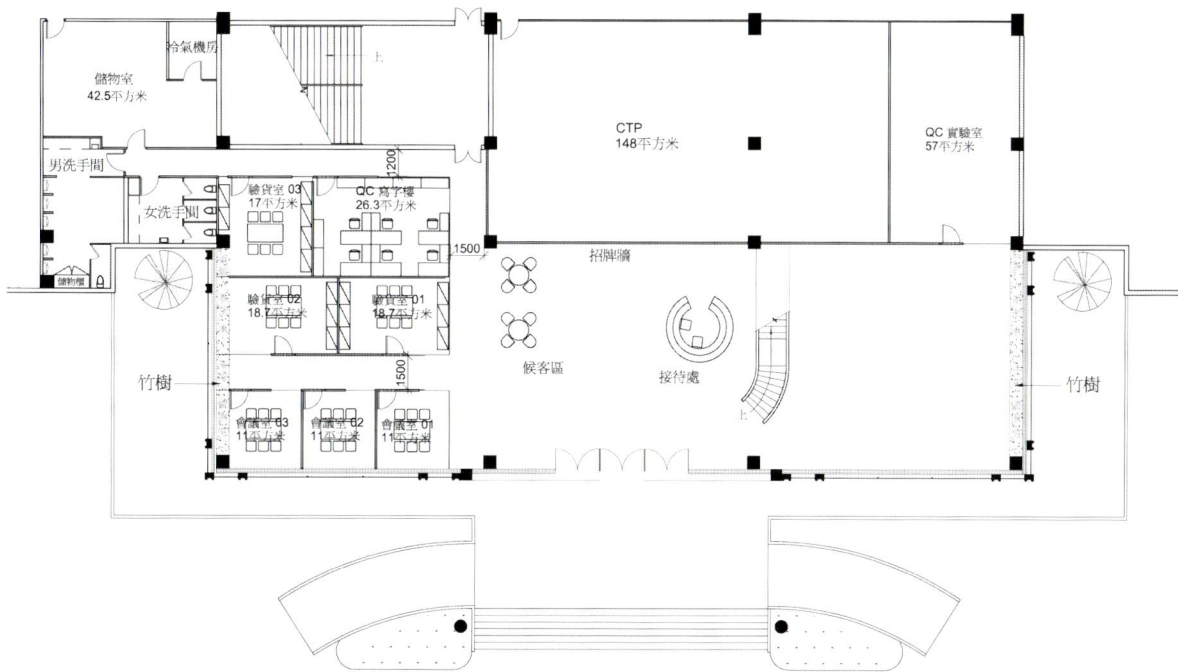

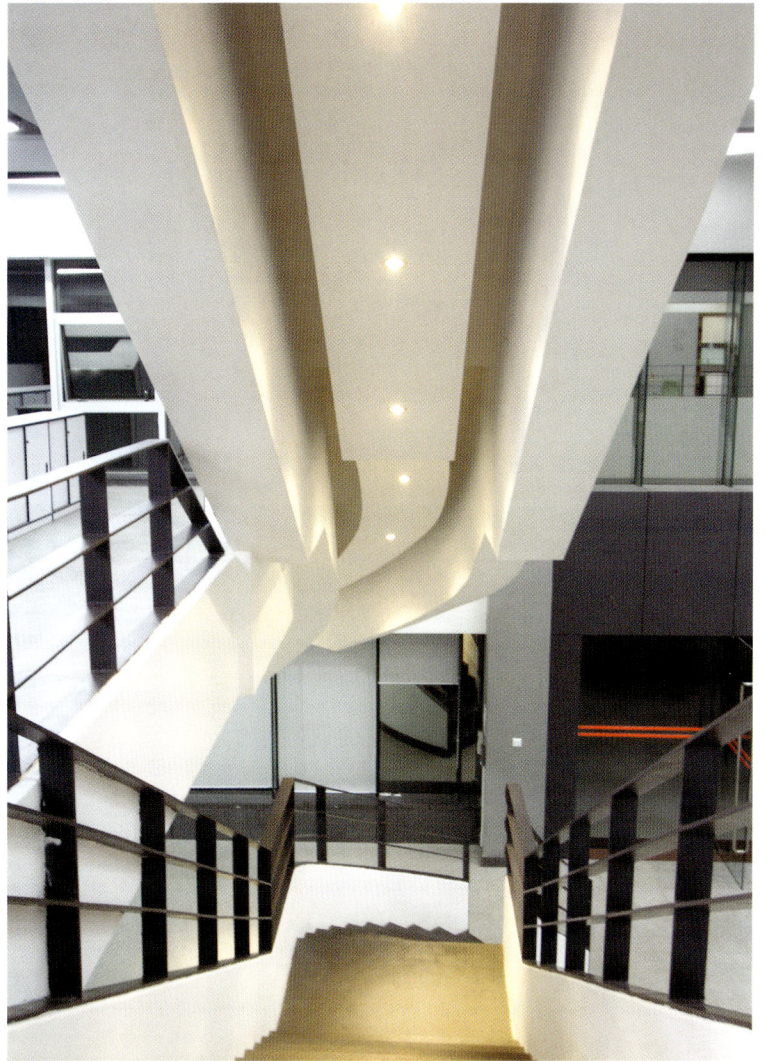

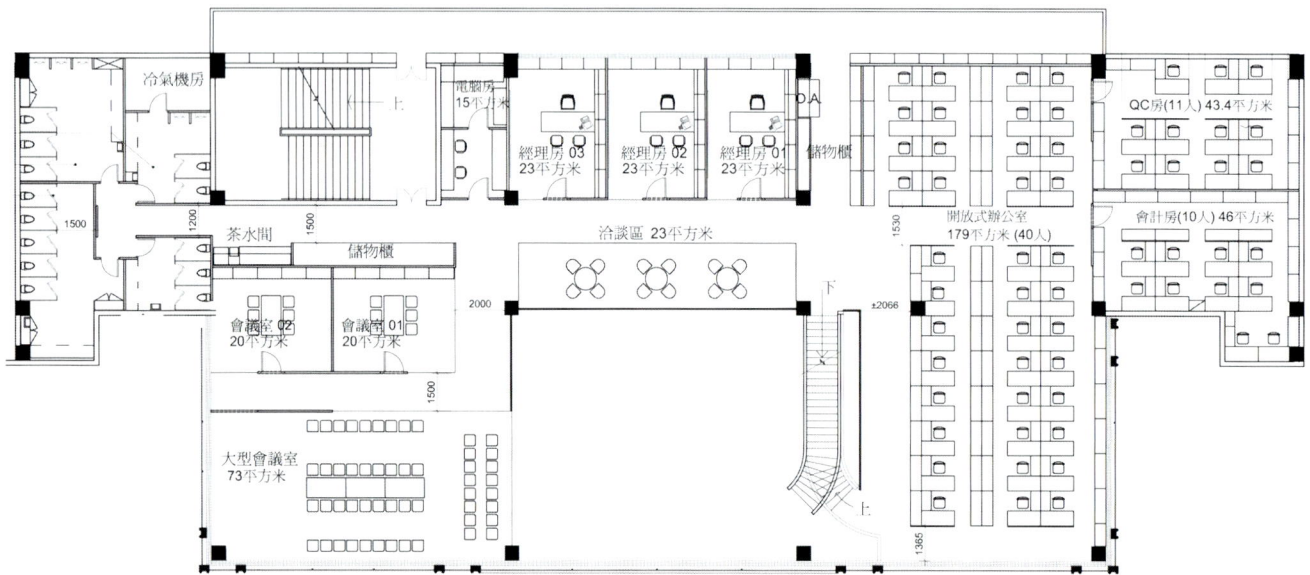

冷氣機房

電腦房
15平方米

經理房 03
23平方米

經理房 02
23平方米

經理房 01
23平方米

儲物櫃

QC房(11人) 43.4平方米

茶水間

儲物櫃

洽談區 23平方米

開放式辦公室
179平方米 (40人)

會計房(10人) 46平方米

會議室 02
20平方米

會議室 01
20平方米

大型會議室
73平方米

1500

1200

1500

2000

1500

1530

±2066

1365

上

下

上

OA

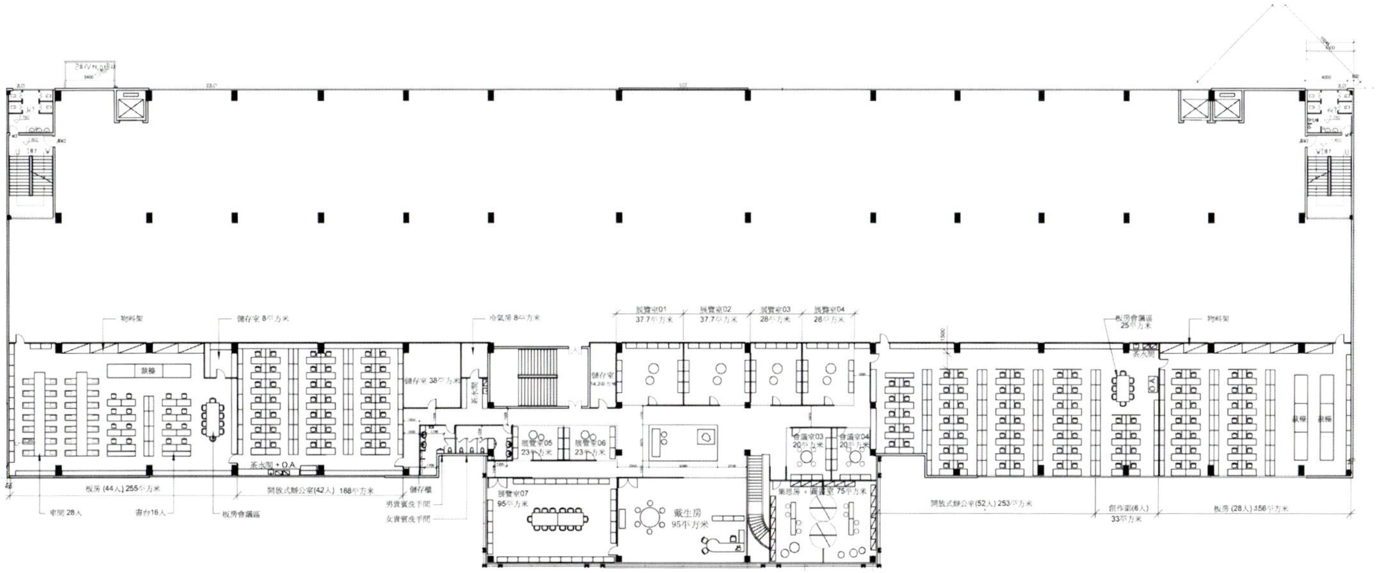

物料架
儲存室 8平方米
冷氣房 8平方米
展覽室01
37.7平方米
展覽室02
37.7平方米
展覽室03
28平方米
展覽室04
28平方米
板房會議區
25平方米
物料架
儲存室 38平方米
茶水間
儲存室
14.2平方米
展覽室05
23平方米
展覽室06
23平方米
會議室03
20平方米
會議室04
20平方米
茶水間 + O.A.
O.A.
板房 (44人) 255平方米
開放式辦公室(42人) 168平方米
儲存櫃
男貴賓洗手間
女貴賓洗手間
車間 28人
書台16人
板房會議區
展覽室07
95平方米
戴生房
95平方米
集思房 + 圖書室 75平方米
開放式辦公室(52人) 253平方米
創作部(6人)
33平方米
板房 (28人) 156平方米

Artapower New Office in Shenzhen

万代华艺美文化用品公司（深圳）

地　　点：深圳
面　　积：3200平方英尺
设 计 师：蔡明治
设计单位：蔡明治设计有限公司

Artapower's office in Shenzhen has more sense of space than its Hong Kong office. Several sections are planned by taking advantage of the characteristics of the original factory building. The first floor is for the reception section, customer waiting section and staff working section, as well as the large and small conference rooms. The second floor is for the workshop and sample section. The top floor is the exhibition room and tea room for visitors.

Artapower设于深圳的办公室及厂房比香港的办公室更有空间感。利用厂房原有的建筑特质，规划出不同部门的分区。首层为接待区、候客区、员工工作区、大小型会议室等。二层部分为厂房及样品区。顶层主要为招待客人参观的产品展览室及茶水间。

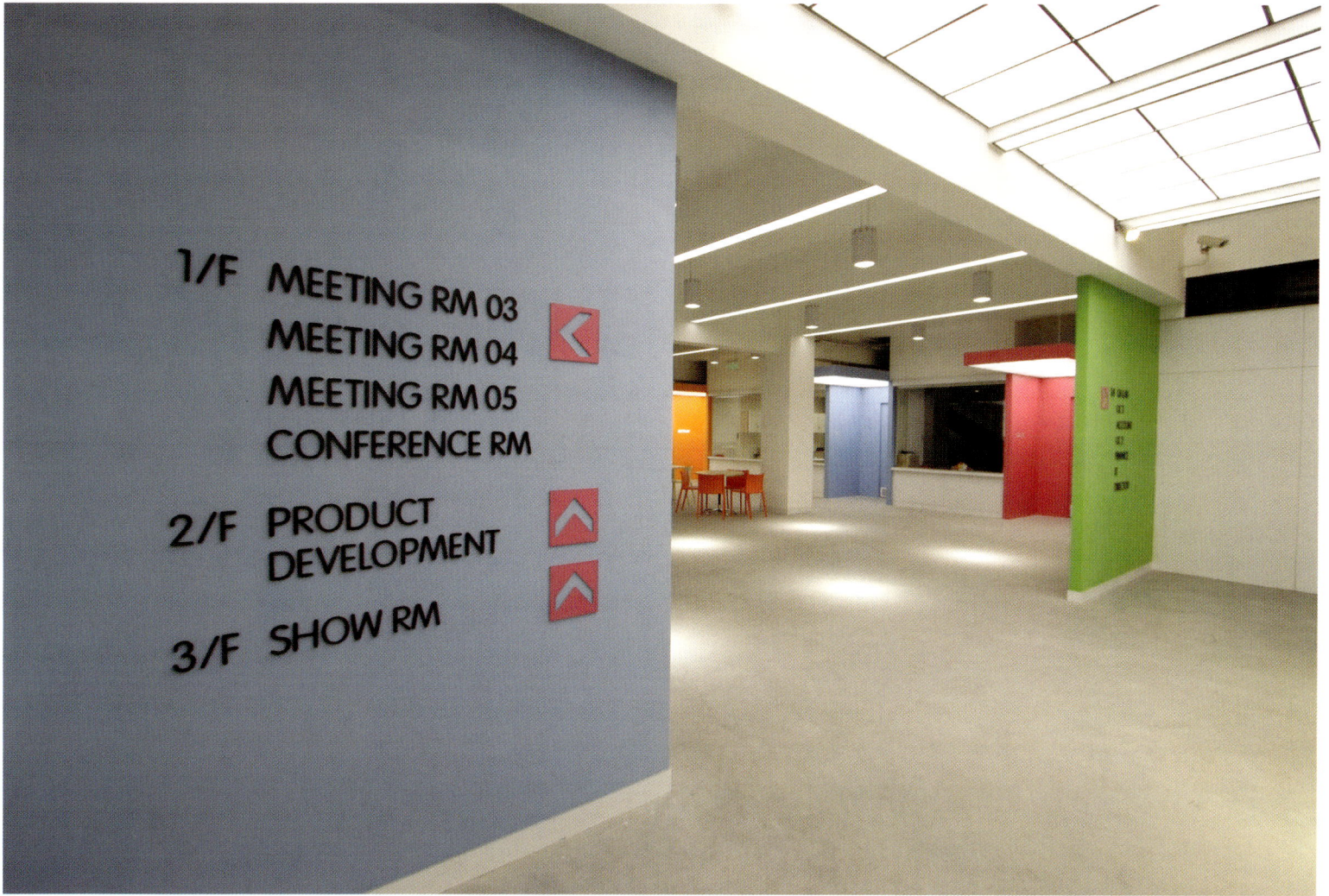

1/F MEETING RM 03
     MEETING RM 04
     MEETING RM 05
     CONFERENCE RM

2/F PRODUCT
     DEVELOPMENT

3/F SHOW RM

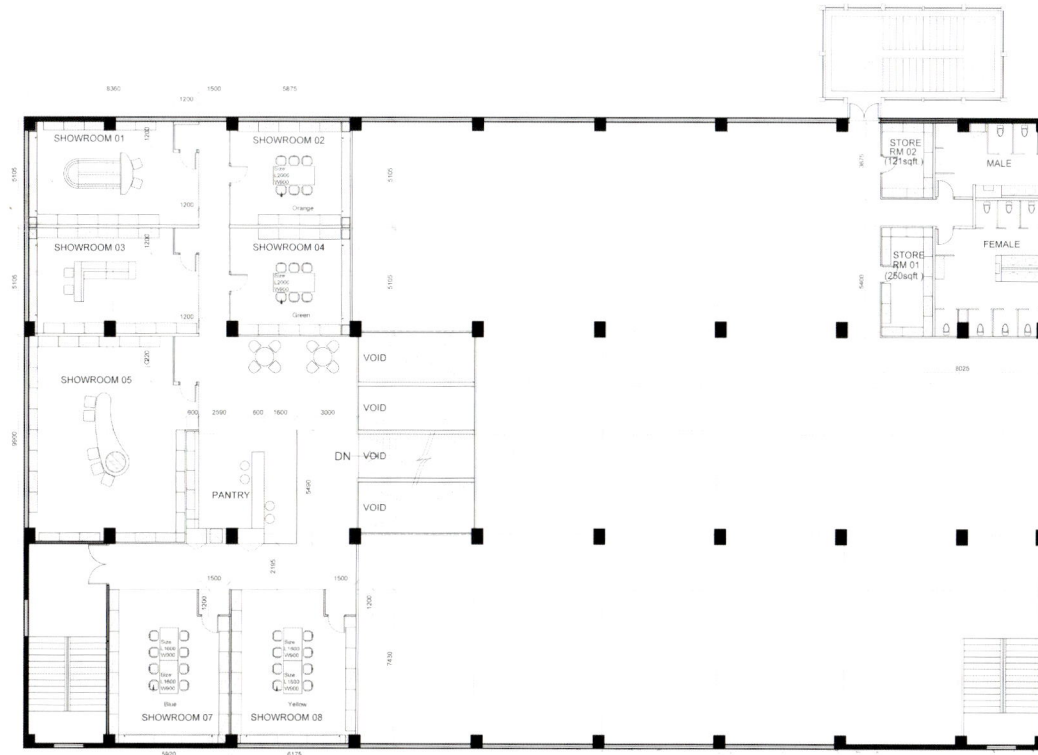

SHOWROOM 01

SHOWROOM 02
Orange

SHOWROOM 03

SHOWROOM 04
Green

SHOWROOM 05

PANTRY

VOID

VOID

DN   VOID

VOID

STORE
RM 02
(121sqft)

MALE

STORE
RM 01
(250sqft)

FEMALE

SHOWROOM 07
Blue

SHOWROOM 08
Yellow

QA LAB

1/F QA
QC
AC
QC
FIN
IT
DIR

FINAN

SHOW RM 05

Artapower New Office in Hong Kong

万代华艺美文化用品公司（香港）

地　　点：香港
面　　积：15000平方英尺
设 计 师：蔡明治
设计单位：蔡明治设计有限公司

Artapower is a company that designs and manufactures products. Different from ordinary manufacturers, it also provides products design for customers. Its office in Tsuen Wan, Hong Kong, is an old-styled factory building. The upper floor of this two-storied office building is mainly for staff working area, reception area, meeting room and the collective thinking room exclusively for CEO and designers. The whole design matches with the overall image of Artapower. The round dark gray steel stairs that connect the two floors are the focus of the whole office. From the stairs, we can walk downstairs which is placed for serving visiting customers as well as the tea room and products exhibition rooms of small and large sizes. The lighting of the public area of this floor is set dimly on purpose, drawing the theme back to the exhibition room.

Artapower是一间产品设计及生产的公司。与一般生产商不同的是，其也为顾客提供产品设计的服务。办公室位于香港荃湾区，是较旧式的工厂大厦。办公室占地两层。上层主要是职员办公区、接待区、会议室、总监和设计师专用的集思房。整个设计要配合Artapower公司的整体形象。深灰色圆形钢铁楼梯贯通了上下两层，亦是整个办公室的焦点所在。穿过楼梯下层主要是用做招待客人参观的地区，分别有茶水间及大大小小的产品展览室。这层公用地方的位置，灯光刻意调得比较暗，令主题回到展览室上。

ARTAPOWER INTERNATIONAL
GROUP LIMITED
藝達堡集團有限公司

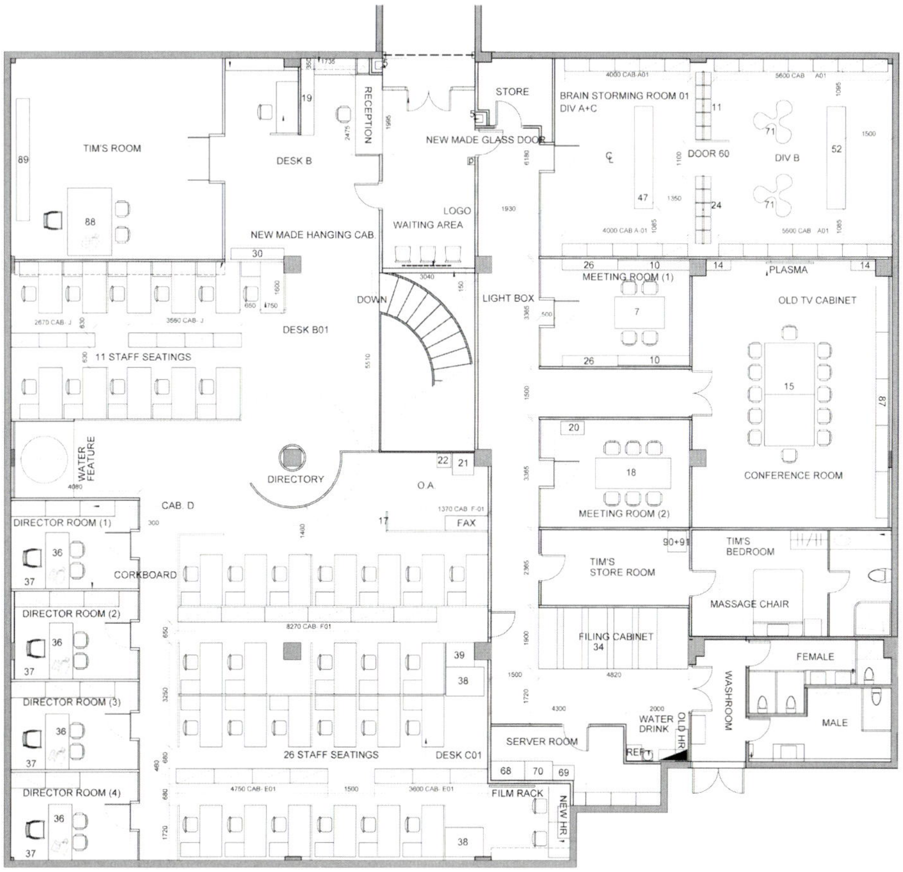
TIM'S ROOM
DESK B
RECEPTION
STORE
NEW MADE GLASS DOOR
BRAIN STORMING ROOM 01
DIV A+C
DOOR 60
DIV B
LOGO
WAITING AREA
NEW MADE HANGING CAB.
MEETING ROOM (1)
PLASMA
OLD TV CABINET
DOWN
LIGHT BOX
DESK B01
11 STAFF SEATINGS
WATER FEATURE
DIRECTORY
O A
CONFERENCE ROOM
MEETING ROOM (2)
CAB. D
FAX
DIRECTOR ROOM (1)
CORKBOARD
TIM'S STORE ROOM
TIM'S BEDROOM
MASSAGE CHAIR
DIRECTOR ROOM (2)
FILING CABINET
FEMALE
WASHROOM
DIRECTOR ROOM (3)
MALE
WATER DRINK
OLD HR
SERVER ROOM
REF
26 STAFF SEATINGS
DESK C01
DIRECTOR ROOM (4)
FILM RACK
NEW HR

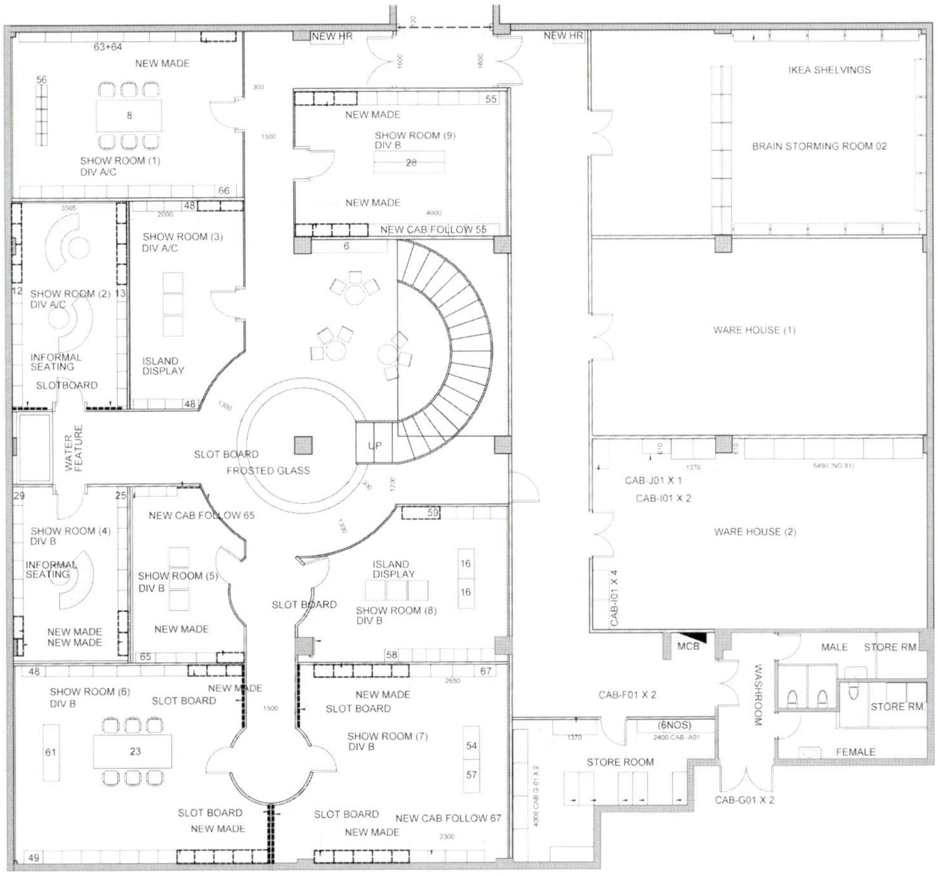

Floor plan labels:

63+64, NEW MADE, NEW HR, NEW HR, IKEA SHELVINGS
56, SHOW ROOM (1) DIV A/C, NEW MADE, SHOW ROOM (9) DIV B, BRAIN STORMING ROOM 02
66, NEW MADE, 55
SHOW ROOM (3) DIV A/C, NEW MADE, NEW CAB FOLLOW 55
SHOW ROOM (2) DIV A/C, INFORMAL SEATING, SLOTBOARD, ISLAND DISPLAY, WARE HOUSE (1)
WATER FEATURE, SLOT BOARD, FROSTED GLASS, UP, CAB-J01 X 1, CAB-I01 X 2
SHOW ROOM (4) DIV B, NEW CAB FOLLOW 65, WARE HOUSE (2)
INFORMAL SEATING, SHOW ROOM (5) DIV B, ISLAND DISPLAY, SHOW ROOM (8) DIV B, CAB-I01 X 4
NEW MADE, NEW MADE, NEW MADE, SLOT BOARD, MCB, MALE, STORE RM
SHOW ROOM (6) DIV B, SLOT BOARD, NEW MADE, SLOT BOARD, CAB-F01 X 2, STORE RM
61, 23, SHOW ROOM (7) DIV B, 54, 57, FEMALE
SLOT BOARD, NEW MADE, SLOT BOARD, NEW CAB FOLLOW 67, NEW MADE, STORE ROOM, (6NOS), CAB-G01 X 2, WASHROOM

ARTAPOWER INTERNATIONAL
GROUP LIMITED
藝達堡集團有限公司

DIRECTO

OFFICE A

CREATIVE
PURCHAS
OPERATI
DIVISION
DIVISION

CONFERENCE RM

MEETING RM 02

Creation Design & Dongtong Construction Office

缔造组设计&东通制造办公室

地　　点：武汉
面　　积：约70平方米
设 计 师：赵国华
设计单位：武汉缔造组装饰设计工程有限公司

This interior design company is located in Room 311 of Wuhan theatre, covering an area of about 70 square meters.

From the air conditioner to the wall, the partition space, the furniture and the furnishings, the creamy white color is the key color to decorate the office. Red and black chairs and shelves decorate the office with original warmth and the simple modern style. The entrance, the working hall, the meeting room, and the manager's room are reasonably divided and full of sense of order. The huge glass board for the meeting table is placed on the massive X-shape crossed table legs. Through the transparent glass, the four lines of "clone legs" give people a sense of rhythm. The connected inverted V-shape footplate not only connects the table legs and stabilizes the glass table board, but also relaxes people's feet when they want to have a rest.

The use of Northeast China ash wood matte partition wall, the carpet, the quardrant iron plastic spraying and the matte fir materials makes people feel quite comfortable.

One thing that is worth mentioning is the design of entrance-two beams of spot lamps' oval-shaped light spot clearly mappes the company's plaque and the sand spraying brand on the glass door which fully expresses the character of design. On the wall beside the door, there is a line shape which is drawn by the designer with a pencil expressing the designer's beautiful intention. The line integrates with the sign of the creation group's design and the sculpture of Terra Cotta Warriors in the contrastive atmosphere of yellow and gray obviously—finding their expressions in the classic and artistic character of "unearthed cultural relic".

这个室内设计装饰公司位于武汉剧院311室，面积只有70平方米左右。

由奶白的空调的雾色系列出发，一直弥散"传染"到墙面与隔断、家具与陈设。红与黑的坐椅与物品柜架，装点出原始的热烈与现代的单纯。划分合理的入口、工作厅、会议室、经理室，空间排列合理，充满了秩序感。超大尺度的玻璃会议桌平板，安置在厚重的X形交叉的桌子腿上，四排"克隆大腿"，透过玻璃，给人有节奏的韵律感。连接的倒V形踏板既起到了连接桌腿，稳定玻璃桌面的作用，又起到了为坐着的人置腿休闲的功用。

水曲柳亚光隔墙板，地毯，方钢喷塑，亚光木杉格的材料运用，都使人感到十分贴切。

尤其值得一提的是入口的设计，两盏投光灯的两束椭圆形光斑，清晰地映射出公司的牌匾，玻璃门上喷砂的产品标志，彰显出设计选用的品质。入门后的一米宽的墙上设计师手握铅笔画出的线形，表达出设计的美丽图面。它与兵马俑的雕塑，缔造组设计的标识，统一在黄灰色对比的气氛中，彰显出有如"出土文物"般的艺术经典品质。

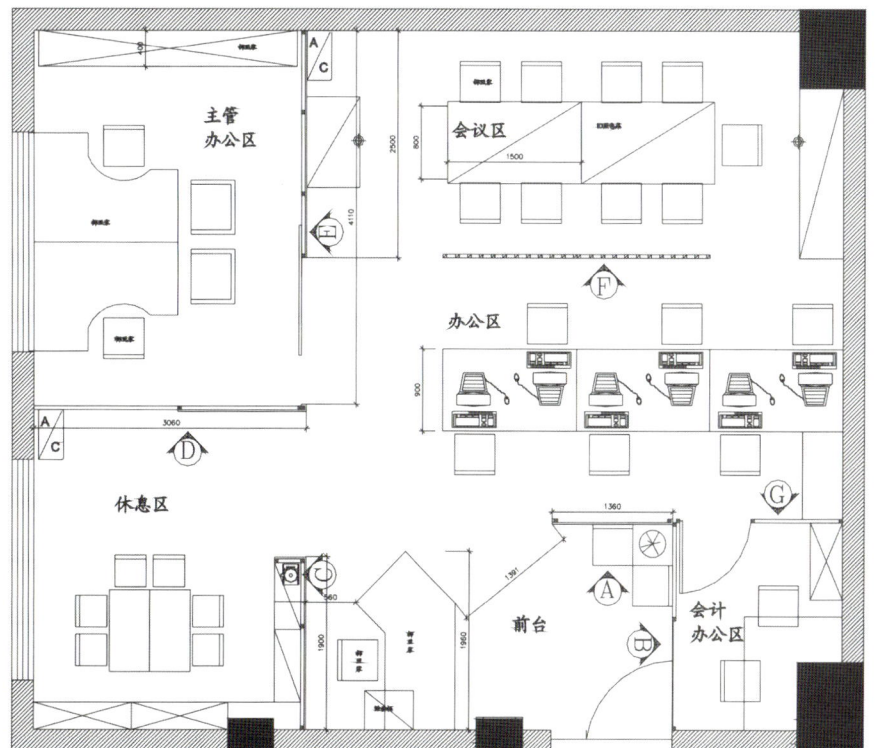

主管
办公区

会议区

办公区

休息区

前台

会计
办公区

Dingcheng International

鼎城国际

地　　点：深圳
面　　积：1100平方米
设 计 师：倪阳
设计单位：深圳市极尚建筑装饰设计工程有限公司

The uncovered fire protection pipeline is painted in red; the Japanese style partition block is in gray. The transparent and hollow glass bricks are put on the walking floor as the slab floor. The partial arc shaped wall, the big tea table, and the flat-rolled steel armrest extend with the hollow stairs composed only of foot boards. The spark of the design idea bursts in the quiet space, flowing like the music, and makes the artistic charm fixed in the various interpretations.

Here one can experience the wise and the emotion of the space, the modern flowing, transparent, capacious, leisure, interesting and changeable office environment; one can feel the space's material perfection, the extending and exquisite quality and the convenient office serving system. The design style will make the user have a different personal experience.

Living in the uproarious environment, but having elegant garden building and oasis-like mini garden with viewing compass, and enjoying the natural and comfortable mood, one will make his tensional body enjoy the relaxation and relief of the nature.

明露的消防管道被涂上了红色，一向被用来作为日式的木本色隔断被涂成了灰色，透明的空心玻璃砖被放置在行走的地面当做楼板来使用。局部的上下弧形墙面，办公桌尺度的茶几，扁钢的扶手顺着只有踏板的空透楼梯延伸。设计思想的火花绽放在凝固静谧的空间中，如同音乐般地自由流淌，让丰富的艺术魅力定格在丰富的解读之中。

体验空间的理智，体验空间的情感，体验现代流动、透明、开敞、休闲、趣味且多样变化的办公环境，体验空间材质的精湛、舒展与细腻，体验便捷配套的办公流程服务系统。体验设计会让使用者有不同的个性体验。

身处喧嚣都市的纷繁环境，赢得幽雅的楼层花园及绿洲般的观景平台的微型小园林，品味自然舒缓的心情，会让紧张的身体得以本性的回归、松弛与释放。

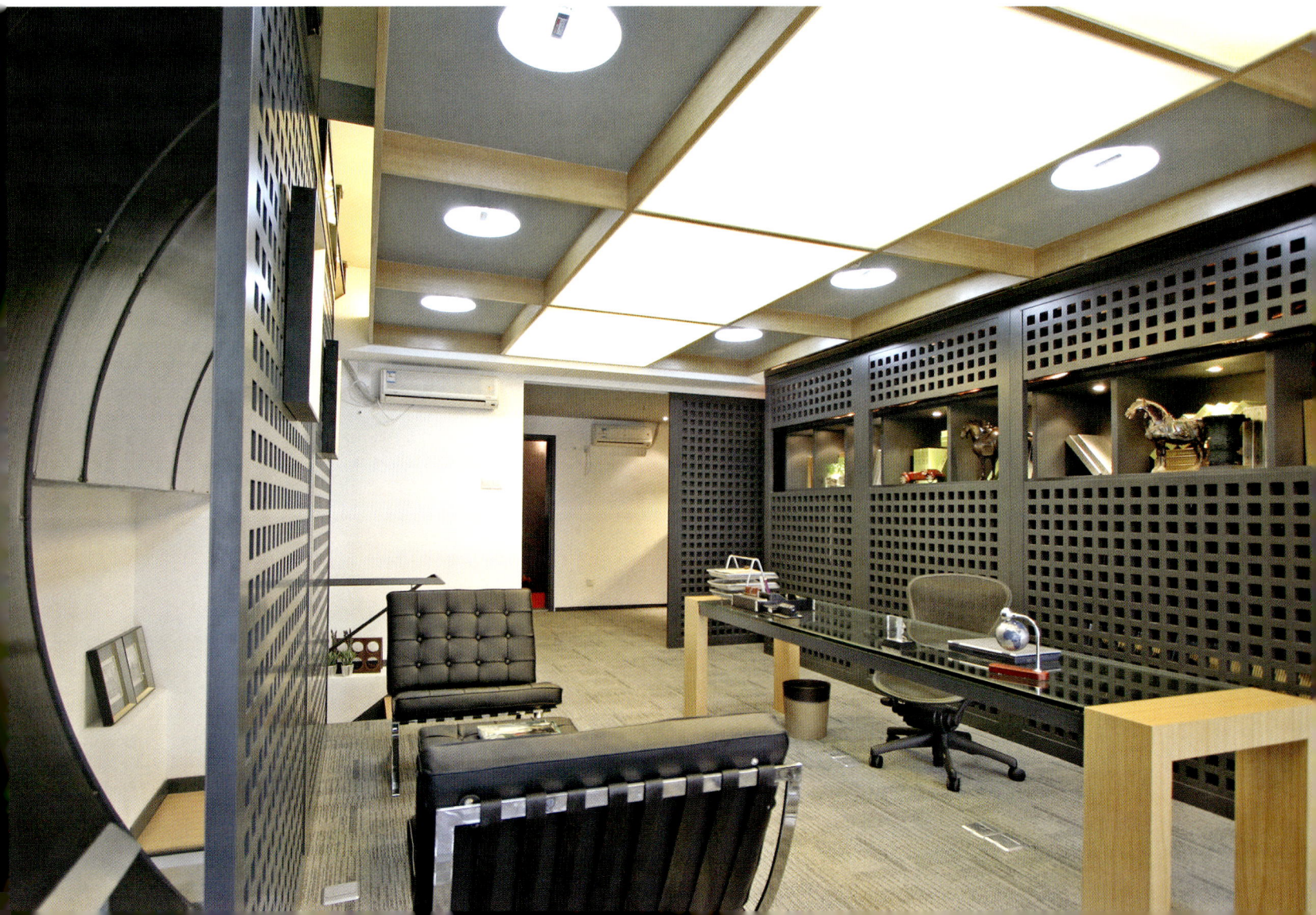

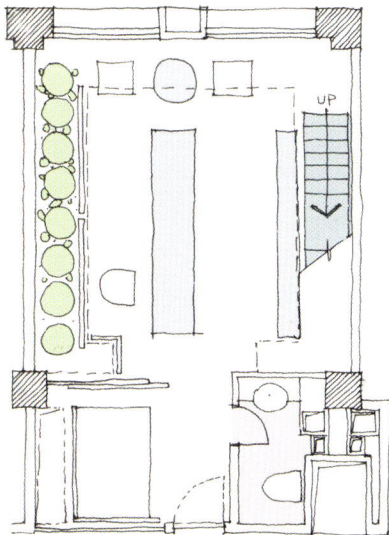
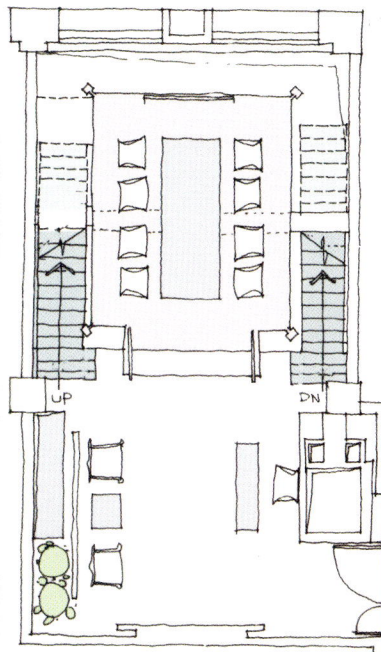
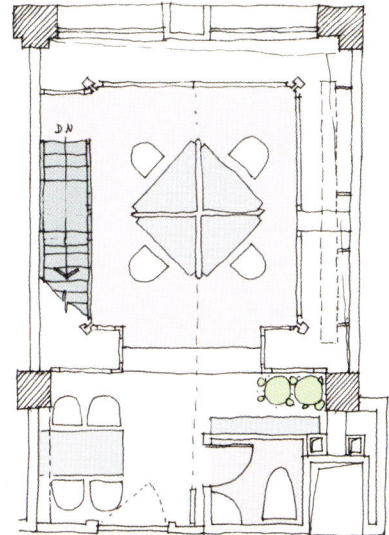

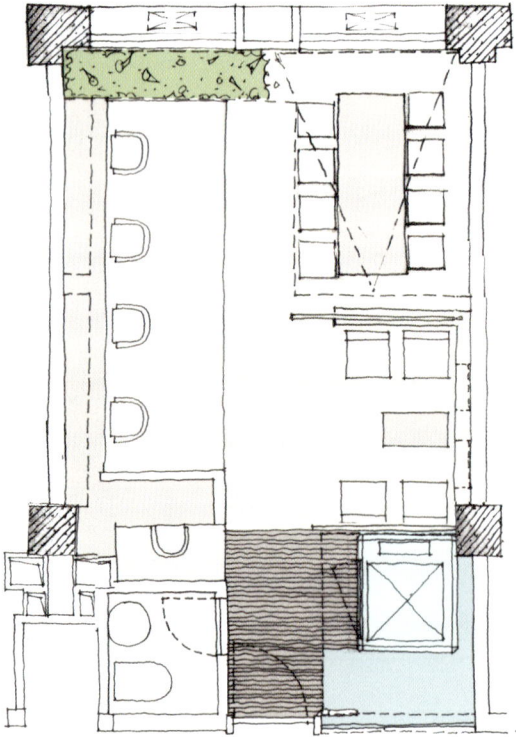

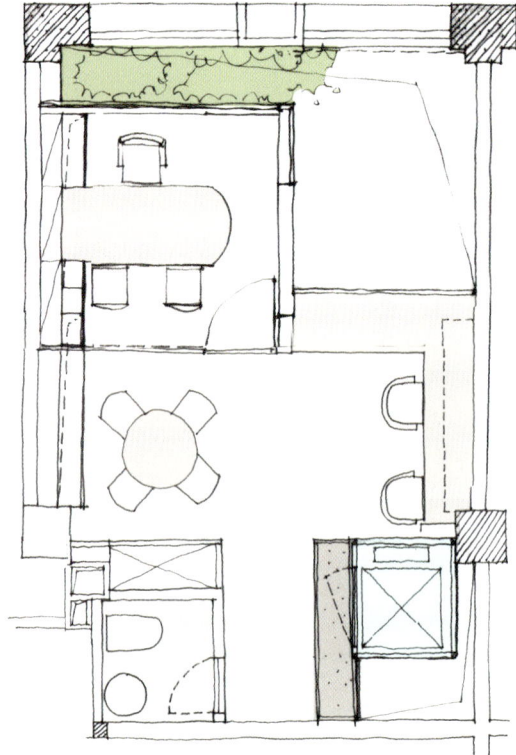

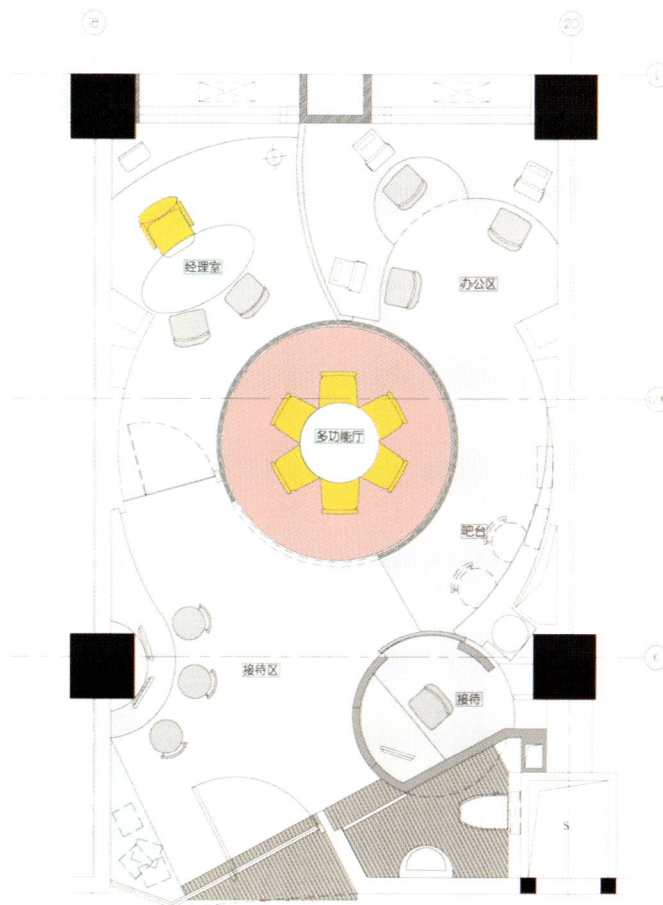

经理室

办公区

多功能厅

吧台

接待区

接待

The Reception Section of Pin Chuan Design Co., Ltd.

品川设计顾问有限公司客户接待区

地　　点：福州
面　　积：180平方米
设 计 师：张慧勇　周华美　郭继
设计单位：福州品川设计顾问有限公司

A black mirror wall at the foyer reduces the narrowness of the space at once. A sunk square area at the right hand side is filled with white stones, and planted in it a branch of tree which is also painted in white. The natural elegance is restored and kept in the office building. The form of the space is shaped by fitment, but the relationship between spaces is kept by decoration. In the use of materials, designers break the bound of interior and exterior rooms in an explorative way. They move these exterior materials inside, and this stroke of genius is like a gentle breeze bringing charm and poetic romance into the space.

To produce a sensible space effect, many furnishing ornaments with a strong sense of design are selected by designers. With the lighting effects from the lamps and lanterns, they achieve a wonderful match with each other, and enforce the characteristics of the space and create a sense of integrity. For instance, several Chinese-style painting frames on the wall form a contrasting match with the luxurious sofa, end table and droplight in the center of the hall: this is original but natural. The visual contrast caused by the materials is unconventional yet not unreasonable, which is just the special feeling that this space brings to us.

入门的玄关处用了一面黑镜玻璃墙，顿时缓和了空间的局促。右手边一个凹陷的方形区域被填上了白色的石块，上面种着一株同样被漆成了白色的树丫，自然的精致被还原在了写字楼里。空间的形式需要通过装修来形成，而空间的关系则要通过装饰来维持。在材质的使用上，设计师大胆地打破了室内和室外的限制，将这些室外的材质移入到室内，成为空间里的神来之笔，仿佛一缕清风为空间带来了情趣和意境。

为了营造充满灵性的空间效果，设计师选择了许多设计感强的家具饰品，并用融于其中的灯具所产生的光线效果实现了相互间的美妙搭配，强化了空间特征，让空间具有了整体性。例如，设计师在房间内的墙上挂了几个中式画格，与大厅中央颇具奢华感的沙发、茶几、吊灯对比搭配，新颖别致又理所当然。这种材质产生的视觉反差，标新立异却也不离谱，这正是这个空间带给人的特别感觉。

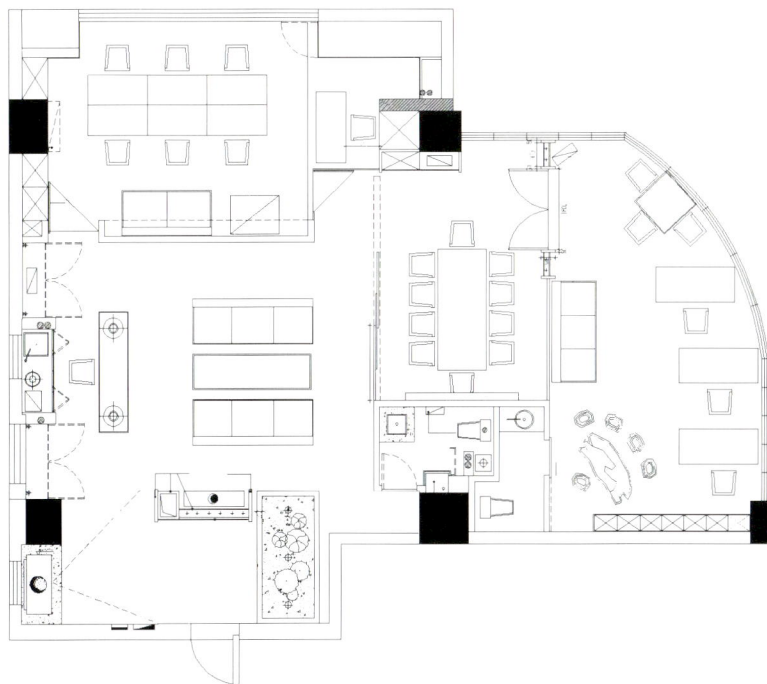

To design is to show. A magic player is the main character of a magic show. Even the most amazing magic show still needs audience's encouragement. If we see an exhibition design project as a magic show, then the designer should be the "magic player" while the visitors are the audience.

F FW Office
福房网办公室

地　　点：福州
面　　积：360平方米
设 计 师：郭继
设计单位：福州品川设计顾问有限公司

Red is full of passion and unrestraint which has an adolescent energy all the time. It seems that any quiet rest has never occurred. Because it is clinging, it is brilliant. When it meets this quiet and lucky space, it brings energy. What the space gives it is a kind of comprehension. Breathing with each other, it seems so fresh and natural. The dynamic red goes to every corner of the space. It seems to be telling people the value of human life. And the lucky and quiet state tells us: have a rest when it is proper; think and recall because it is a short cut to improve by review and learning. There is a group of people who are full of energy as well. They are trying to create the value of life. So they get together and compose the movement of life together.

"红色"热情、奔放，永远都充满了青春的活力，似乎没想过要安静地停歇一下。因为执着，所以灿烂。当它与这个安静、祥和的空间相遇时，为其带来了一份活力，而空间赋予它一份包容。相互呼吸着，如此清新、自然。活跃的红色穿梭于空间每个角落，似乎在告诉人们生命的价值。那份安静祥和则说：适当的时候应该停下来休息了，顺便思考并回忆一下，因为温故而知新也是进步的捷径。

有那么一群人，他们也充满了青春活力，并且在努力创造着生命的价值。于是，他们与它们便走到了一起，共同谱写这首生命的乐章。

Boutique Office at Central

中央精品办公室

地　　点：香港
设 计 师：何周礼
设计单位：何周礼建筑师有限公司

Boutique-style Gallery Spine

Instead of conventional reception lobby, the prologue of this office is a Boutique-style gallery which is reserved for multi-functions such as exhibitions and cocktail events.

The Gallery is also equipped with multi-lighting levels catered for different events, and audio-visual equipment is installed.

While conventional lobbies are shiny and bright for a sense of grandeur and class, this Office takes another approach to convey a sense of style with black as the main colour. The inclined Black-Textural Tile Wall intensifies the strong-sense perspective towards a reception counter, where Asian style settings are placed as a waiting area.

Directors' Chamber

Ignoring the conventional programming and typology of Directors' Rooms, a Hall of Fame is designated as the Second Spine for highlighting the projects for visitors.

The Directors' Rooms are thus conceived as "Glass Boxes" along the Hall of Fame and are elevated.

In between the two elevated "Glass Boxes", a VIP lounge performs as an accented space for visitors.

Centre Pond and Open Bar

Designers' working area is circled off in an unconventional elevated ring. Raised platform and overhead shelves offer storage without occupying additional space, thus allow maximized working area for designers. Choice of white is used for the open bar to signify a break from the rest of the working place which is in black and beige.

精品门廊立柱

不同于传统的接待大堂，办公室的进门处是一个精巧的门廊，该门廊是多功能的，例如可以用于展览，也可以举办鸡尾酒会。

为了适应不同的场合需要，该门廊也配备了不同层次的灯光效果和视听装备。

传统的大厅都是灯火通明的，通常带给人们一种庄严感和阶级感。该办公室用另一种方法来诠释着一种时尚感，即用黑色作为主色调。墙体采用黑色材质的瓷砖倾斜铺制而成，有助于加强接待处的透视感，接待处的装饰被设计成了亚洲风格。

董事室

抛开传统的董事办公室的设计和类型，一间名人堂被安排作为第二个中心地带以突出整个项目，供客人使用。

沿着名人堂，董事办公室因此就被视为"玻璃盒子"，并将其位置提高。

在两个被提高的"玻璃盒子"之间，是一间贵宾休息室，它作为一个具有特色的空间供客人使用。

中央大堂以及开放式酒吧

设计师的工作区域被分隔在一个非同寻常的吊高的圆形物之内。凸出的平台以及天花板上的架子提供了储藏空间，并且不需要占用多余空间，这样就为设计师提供了最大的工作空间。开放式酒吧选用白色色调，以区别于其余的以黑色和浅褐色为主的工作区域。

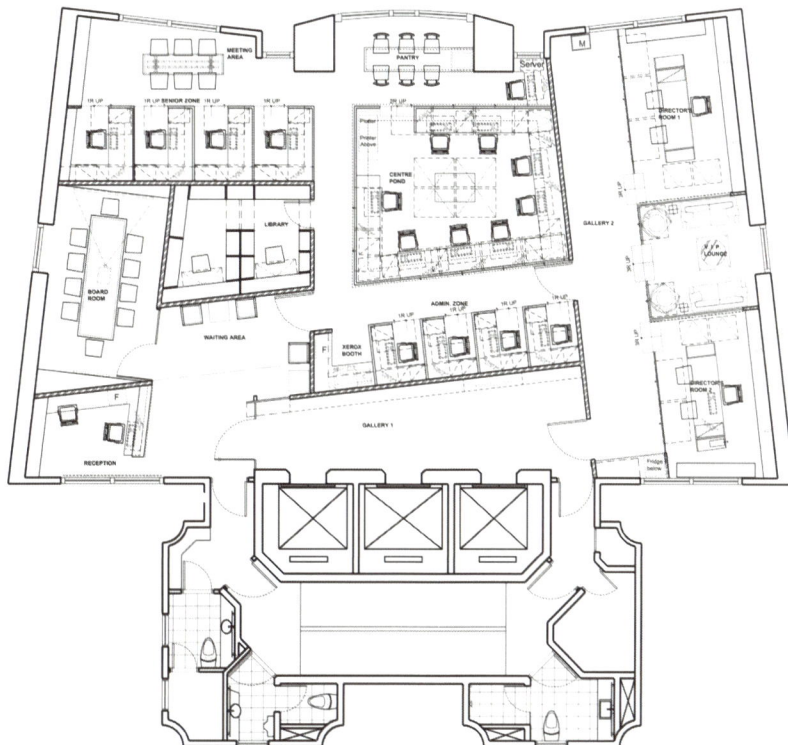

GALAXY

佳威世纪科技有限公司

地　　点：深圳
面　　积：1080平方米
设 计 师：陈飞杰
设计单位：深圳汇杰室内设计工程有限公司

The inner area of Galaxy is 1,080m². The functional division of the whole space is reasonable，and the structure is compact, which is very simple but decent. Because the company deals with IT industry, thus a lot of new originalities on the design are employed. The designer adapts carbon fibre material faceplate for the antehall. The ceiling and LED light designed on the floor look like various stars on the sky which makes a very strong technical sense to the whole space and it has shown the industrial feature of the company.

Two round palavering rooms are set up on the side of the antehall which have been raised up by glass. Plants are put as interspersions which give the whole space a kind of nimbus.

The designer has paid much attention to the creation of the atmosphere for the working area. The collocation of ceiling, wall, the plants on the floor and the blue color gives a kind of energy to the tensional working environment.

　　佳威世纪科技有限公司室内面积1080平方米，整个空间功能分区合理、结构紧凑、简洁大气。因为公司经营的是IT产业，故在设计上采用了很多新的创意。前厅设计上，采用碳纤维材料面板。天花、地面设计LED灯带，看似满天繁星，使得整个空间科技感非常强烈，突出了公司的产业特色。

　　前厅旁设置了两个玻璃抬高的圆形洽谈室，上下点缀绿化植物，赋予了整个空间灵气。

　　办公区亦注重环境氛围的营造，天花、墙面、地面的植物与天蓝色的空间颜色搭配给紧张的办公环境带来活力。

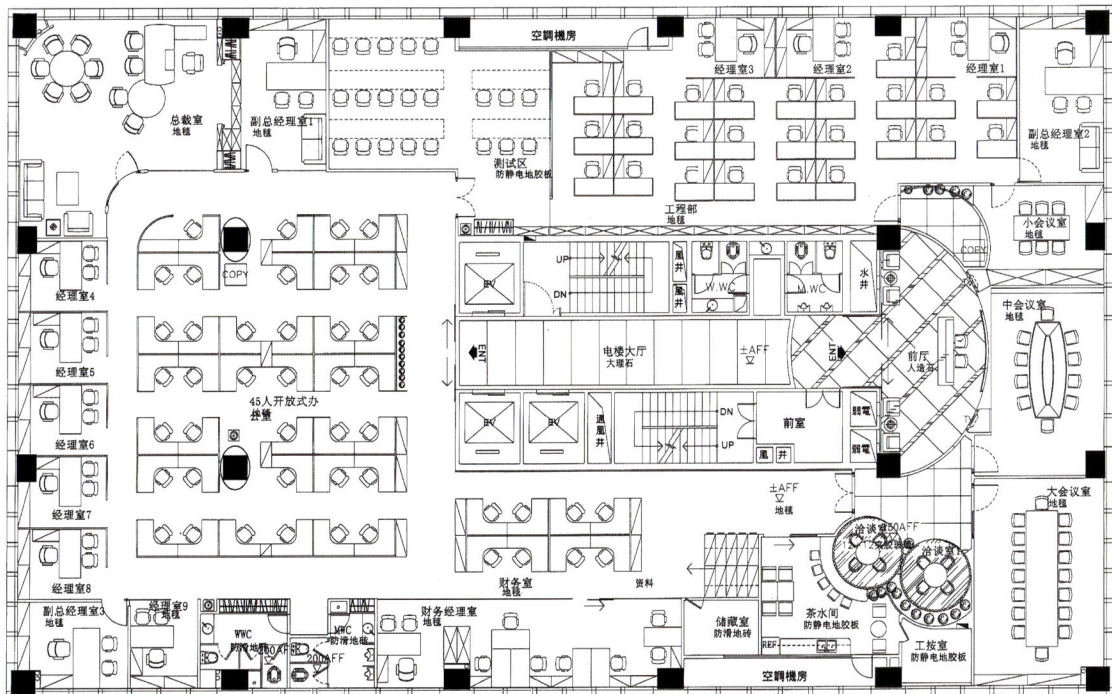

Sir Hudson Office in Kwun Tong

Sir Hudson香港观塘区办公室

地　　点：香港
面　　积：3922平方英尺
设 计 师：蔡明治
设计单位：蔡明治设计有限公司

Sir Hudson, an international food and beverage group, has many brands of restaurants and coffee houses, including Hara Kan, Suzuki, Baan and Teppan. Except for basic facilities, a central kitchen and a place for customers' food tasting are also kept in its new office in Kwun Tong, Hong Kong. There is not much ornament in the whole design, and "being characteristic, unique and aggressive" are what it conveys. No matter it is in the workshop, meeting room or managers' working area, there is enough space for notices or records. No doubt it is a real "working" place indeed.

"Sir Hudson" 是一个国际饮食集团，拥有多个餐厅及咖啡室品牌，包括Hara Kan、Suzuki、Baan、Teppan等等。"Sir Hudson" 位于香港观塘区 (Kwun Tong) 的新办公室，除了一般办公室应有的设施外，还预留了位置设置中央厨房及招待客人试菜的地方。整个设计没有过分的修饰，"性格、个性、冲劲" 是这次设计要表达的信息。无论是在工作间、会议室还是在经理工作区，都留了充足的空间用于告示或记录用途，是一个真正 "工作" 的地方。

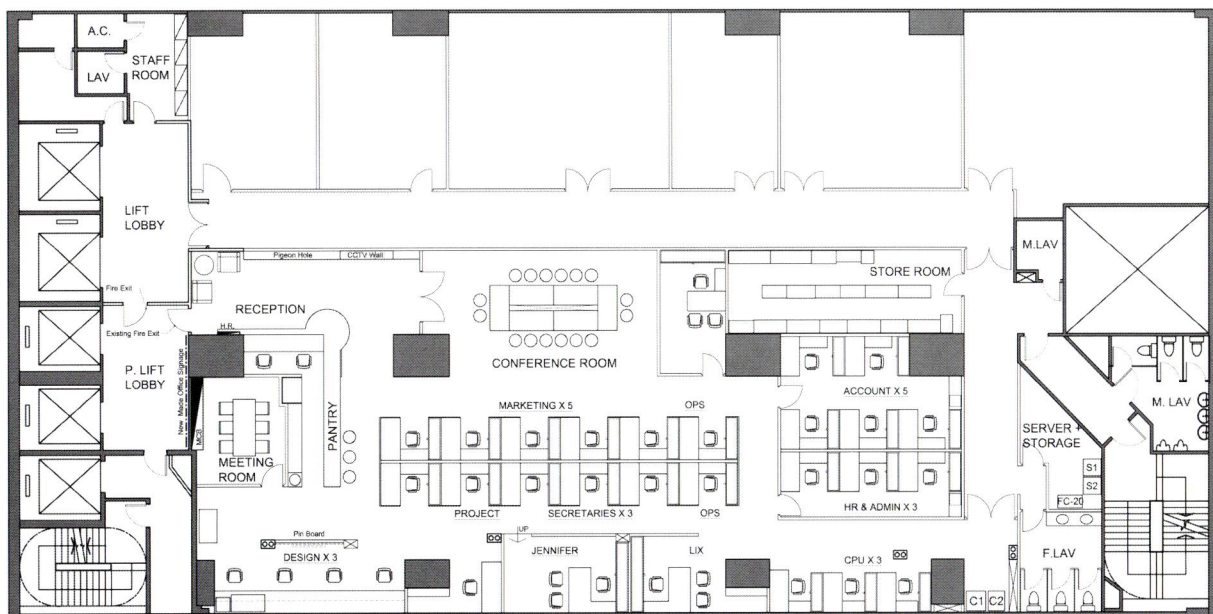

Guangzhou Linguaphone Training Center

广州灵格风培训中心

地　　点：广州
面　　积：约640平方米
设 计 师：郑成标
设计单位：广州成标装饰艺术设计有限公司

The first training center of Linguaphone in Guangzhou is located in the area where business creams swarm. Because of this, the whole style of Linguaphone shows fresh, advanced, fashion and mutual feeling. The whole colors matched are blue, white, silver and red. The space is divided into reception area, hall, movie room, meeting room, training room, electronic mutual place and working places.

The reception place uses white as the main color. The background is fresh and refined green which shows a relaxing, pleasant and fresh atmosphere.

In the hall all things from the floor to the wall or from the wall to the ceiling are all flower patterns made of blue and white mosaic. There is a contrast between the exquisite crystal lamp and rigid mosaic. The red camber sofa increases the gusto of the space. At the same time, it meets the requirement of communication and interaction of the sharing place which forms an advanced place full of fad, science fiction.

The movie room, meeting room and training room go through the design method of fashion, relaxation and advancement which show the advanced, fashionable and modern bardian style of Linguaphone. Together with the open and relaxing teaching feature including movie, music art and chatting, it expresses the transcendent and professional authoritative position among the same occupation.

灵格风首个位于广州的培训中心地处商务精英云集区，正因为如此，灵格风的整体风格展现为清新、前卫、时尚、互动。整体配色为蓝色、白色、银色、红色。区间划分成接待区、大厅、影视室、会议室、培训室、电子互动区、办公室。

接待区以简约干练的白色为主色，背景配以清新脱俗的草绿色，展现出轻松、愉悦、清新的气氛。

大厅由地到墙，由墙至天花都采用蓝、白马赛克拼制成优雅的花状图案。精致的水晶灯与硬实的马赛克形成对比；大红色弧线形沙发为空间增添了趣味，同时也满足了共享空间的交流、互动的要求，形成了一个时尚、科幻的前卫空间。

影视室、会议室、培训室等都是贯穿于整个时尚、轻松、前卫的设计手法之中，展现着灵格风超前、时尚、不失时代步伐的个性风格，配合了灵格风以电影、歌艺、聊天等开放轻松形式教学的特色要求，体现了灵格风在同行中超然、专业的权威地位。

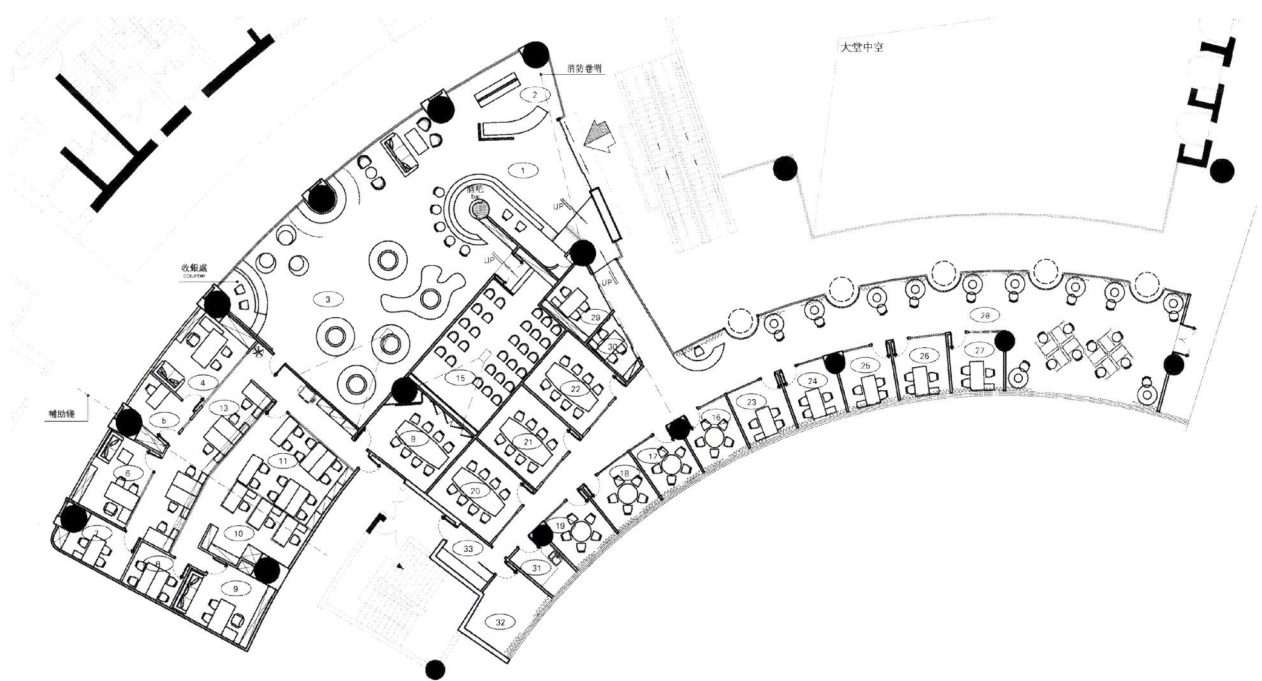

Hong Kong Time×Pose Center Office Building

香港时霸诺德中心写字楼

地　　点：深圳
面　　积：1200平方米
设 计 师：明光华　明罡
设计单位：深圳市写意居装饰设计工程有限公司

As designers of a vogue and unique office building, they emphasize on the color setting: all the interface elements such as grounds, furniture, furnishings and ceilings are all offwhite, which sends out the new and fresh vision impact. Nothing is more suitable to the office than this simple, harmonious, elegant and clear color. It is different from the heavy color and strong contrast in entertainment and business selling places. The calm color brings the clerk proper limits in problem-solving and efficiency in work. Of course the designers don't forget places such as anta corners and chairs' surfaces where the contrastive hue of yellow ocher and amaranthine is used, for it penetrates a little conflict and romance in the wateriness. There is a grass green clapboard of 300mm height and 1500mm length on each connection of the partition and the desktop. So every time the clerk sees this "grassland horizon", his eyes may be accommodated and relaxed. In this circular arc sand grass room, with the decorations of the yellow ocher chair and the flowers on the center of the round table, business negotiation would be much easier.

The apartments' slab, structural girder, lamps and lanterns, aeration equipment and all kinds of routing pipes are all displayed before people's eyes "openly" without any "packaging", which shows the beauty of modern machinery. The contraposition echoes of the elliptical ground to the irradiance ceiling in the reception hall and half elliptical desktop to the half elliptical surface lamp in the assembly room, together with the ceiling vertical plates of the key places and the proper disposal of sculpt, show the designers' perfect control abilities to the space.

作为一个时尚与个性的办公机构，设计师紧紧抓住色彩处理，让整个地面、家具与陈设、天棚等界面要素统一在淡淡的灰白色之中，给人耳目一新的视觉冲击。这种单一、协调、淡雅与寡净的色调最适合办公场所不过了。它有别于色彩浓重、对比强烈的娱乐和商业销售场所。冷静的色彩给办公人员带来处理事务的分寸把握与工作的效率。当然，设计师也没有忘记，在墙柱的角位，坐椅的包面采用了土黄与紫红的对比色调，让平淡中透出一点点冲突与浪漫。每个隔断与桌面连接的300毫米高、1500毫米长的一条草绿色的隔板，使每个工作者的视线可以时常落在这"草场的地平线"上，让疲劳的眼睛不断地得以调节与放松。圆弧形喷砂玻璃的室内，土黄色坐椅与圆桌中心放置的清新淡雅的花卉，让商务谈判轻松了许多。

房间楼板、建筑的梁、灯具、通风设备与各种布线的管道都"光明正大"地展现在视线里，没有做任何的"包装"，给人以现代的机械美。而接待大厅椭圆形地面与发光顶棚，会议室半椭圆的桌面与顶棚的半椭圆面灯的对位呼应，以及重点空间的顶棚垂板与造型的处理，都显示出设计师对于空间把握的能力。

Guangdong Huizhou District Lianhong Landification Exploitation and Investment Co., Ltd. Office

广东惠州联宏石化区开发投资有限公司办公室

地　　点：惠州
面　　积：2000平方米
设　计　师：陈颖
设计单位：深圳秀城设计

In western countries, people are attracted by the wall, but in eastern countries, what attracts people is the floor.

This is a company whose main line is petrochemical industrial park development, exploitation and investment. It is located in Dayawan Economic Development Zone of southern Huizhou, Guangdong Province. It is an eastern company which is eager to get close to the western management style and mode.

A curved line goes into the front and the back halls, so the space can get the effect of extension and flowing. Some local walls which are vertical come inside. And the concave face is divided into many pieces of different sizes. They are combined together like building up toy bricks. The ceiling is full of deal woods which have been hung up by light gage steel joist. So it has the same rhythm as water surface. People will walk on this flow to go to work with a delighted mood.

The simpler the form is, the stronger the space will be.

The intervention of the warm color is not only the extension of the outside color, and the staff here will have such comments: it does not look like an office".

在西方，人们被墙体所吸引，而在东方，吸引人们的则是地面。

该公司是一家以石化工业园区开发建设和投资为主的企业，位于广东省惠州市南部的大亚湾经济技术开发区。它是一个急于向西方的经营管理及模式靠拢的东方企业。

一道弧线介入前后两个厅，空间因而得到延展和流动。作为垂直界面的墙体局部内缩，被凹入的面分割成大小不等的体块，搭积木般地被组合起来。天花被轻钢龙骨吊挂的松木条阵列所充满，有着水面一样的节奏。人们带着愉悦的心情，踏着脚下的流动平面去上班。

形式的语言越单纯，空间就越有力量。

暖色的介入，不仅是外立面颜色的延续，而且可以让员工对这里有这样的评价："看起来不那么像办公室。"

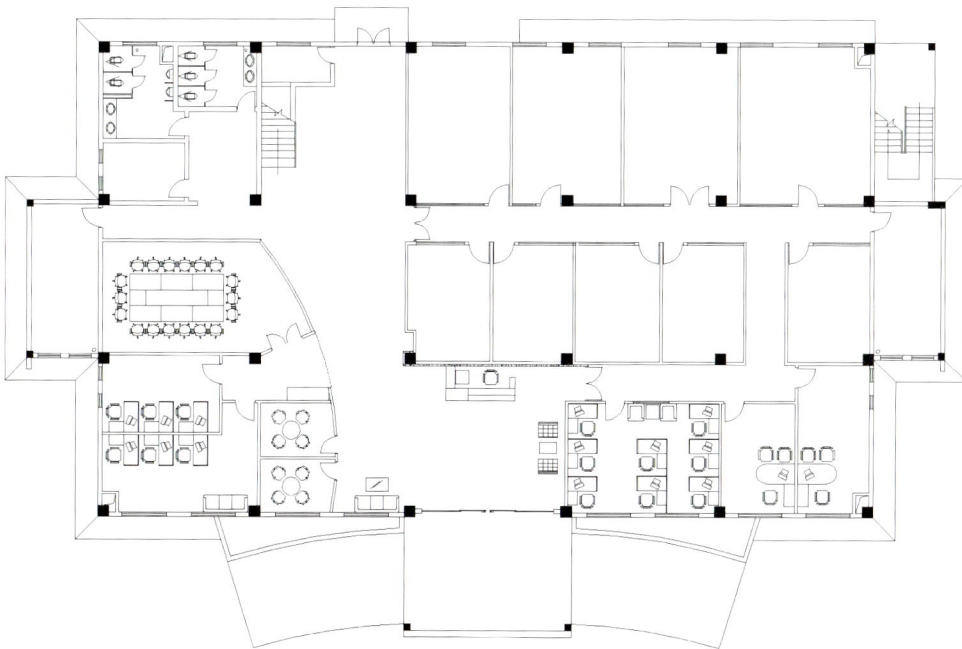

Dickson Hung Design Office

洪德成设计办公室

地　　点：深圳
设 计 师：洪德成
设计单位：洪德成室内设计（香港）有限公司

The designer always tries his best to promote the "brand" when designing a project. He transfers his unique style through the design of office space to attract the guests.

The thin steel wire is employed to hang the heavy section steel and the steel wire net with track lights attached at its bottom for places where local lighting is needed. The black chair and the short black partition on the desk, the black doorframe and picture placeholders, the black drawer faceplate and the decoration of the black English letters of "INTERIOR DESIGN" on the transparent glass partition—all of these have appropriately expressed their unique noble styles.

The simply-designed data tank and abundant collections of books make people learn what the connotation of knowledge is and what the design experience is. The carefully selected teapot, cup, tray, and collected Muhuage, porcelain and bronze make people understand the designer's unique percipient. A group of pictures which are hang on the wall and the huge modern paintings also make people think of the tension of modern art. The simple desk and chair, the elegant sofa and tea table, the charming lamps and the uncanny veneered artificial board will definitely amaze the guests who are coming to sign the contract.

设计师在给自己的团队做设计时，总是倾力打造设计的"牌子"，通过设计公司的空间体验，向客户传递设计师与众不同的品位，以赢得客户的青睐。

天棚用细细的钢丝绳，吊着沉重的型钢与钢丝网，下面附着需要提供局部照明的轨道灯。黑色的坐椅与办公桌的黑色矮隔断、黑色的门框与黑色的图片框、黑色的抽屉面板与透明玻璃隔断上黑色的"INTERIOR DESIGN"（室内设计）英文字母装饰，都在"画龙点睛"地传递出不同寻常的高贵品质。

简洁的设计资料柜，丰富的资料藏书，让人感受到知识的内涵与设计的阅历。精挑细选的茶壶、水杯、托盘，收藏的木花格、瓷器、青铜等艺术品让人窥测到设计师非同寻常的鉴赏能力。而挂在墙上的成组图片与大幅浑厚、充满肌理的现代画品，也让人感染到现代艺术的张力。简洁的办公桌椅，考究的沙发茶几，魅力十足的款款灯具，鬼斧神工的贴面人造板材，都会让前来签约的客商惊叹不已。

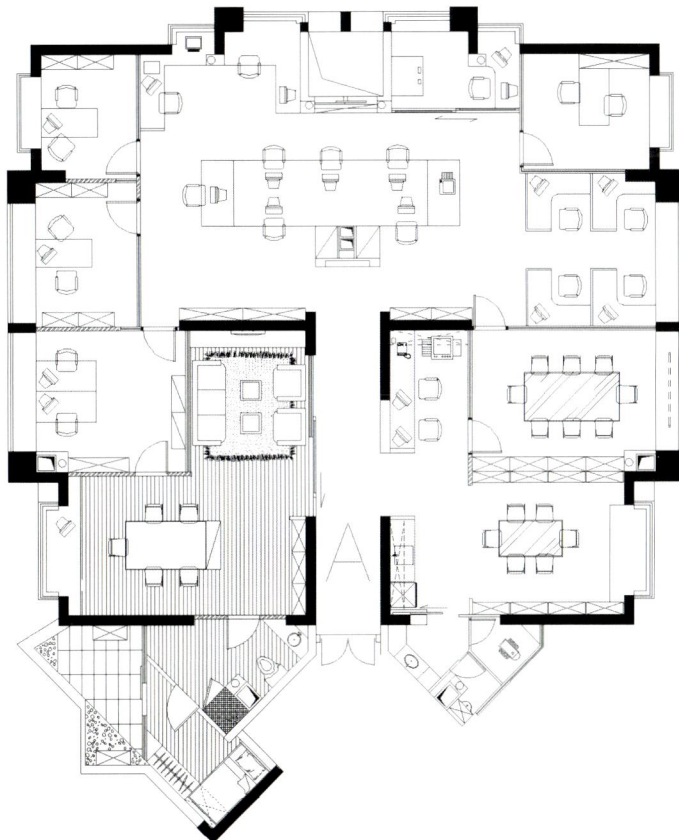

Hongteng Design Company

洪腾设计公司

地　　点：福州
面　　积：300平方米
设 计 师：洪斌
设计单位：洪腾.设计

"Space" Behind the Space

Blending the concept of family into the space will create a cozy atmosphere and sooth our mind. A patch of space will bring us a happy mood all the time.

Everyone who stops to view the project of the Hong Teng Design Co., Ltd. is likely to be infected by the designer's living concept, simple and thorough. When everything is soaked with nicety, humans and the space will be in a harmony.

空间背后的"空间"

把家的概念融入空间，营造出温暖的氛围。人也就不会轻易浮躁。而小小的一方天地更能带来每时每刻晴朗的心情。

在对洪腾设计公司的凝视中，每一个留步的人都容易感染上设计师的生活理念，简单，透彻。当这一切浸润着美好，人与空间便融合在了一起。

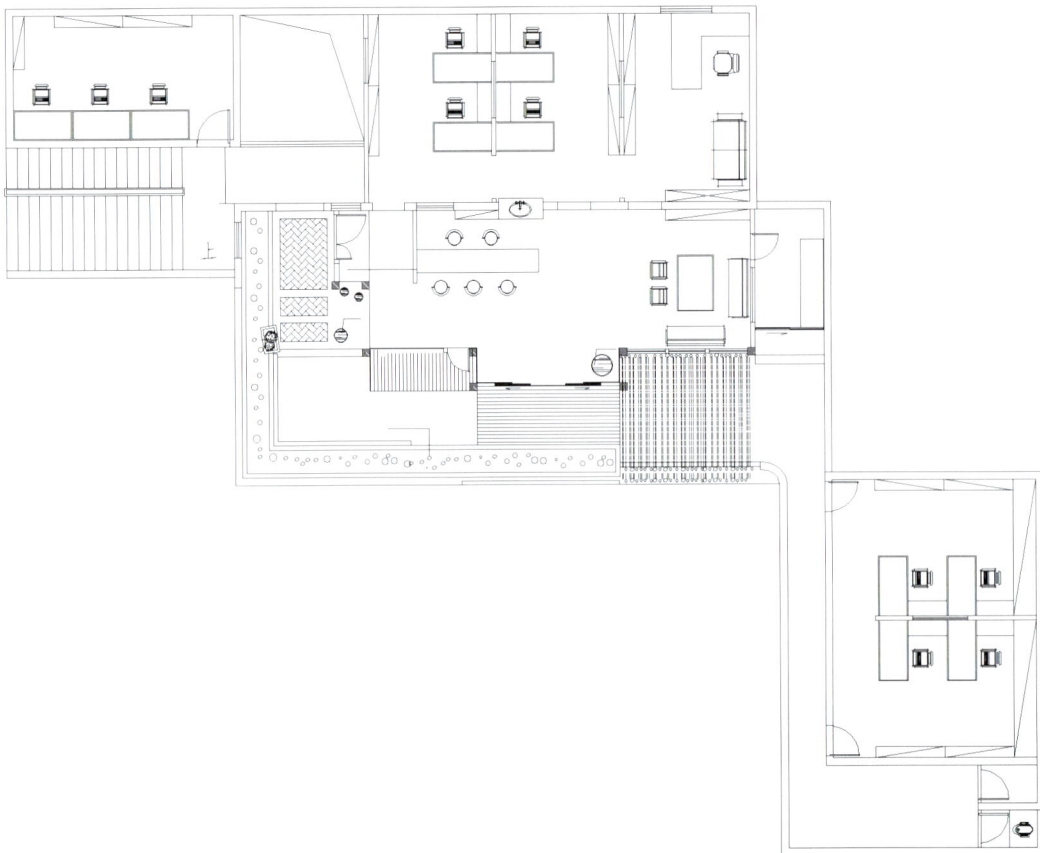

Shenzhen Element Interior Design Ltd. Office

深圳汇杰设计工程有限公司办公室

地　　点：深圳
面　　积：360平方米
设 计 师：陈飞杰
设计单位：深圳汇杰室内设计工程有限公司

For the plan, the designer does not start with the inner space only. He extends the thought to the tangible and intangible multi-dimensional space. Then an exact visual and functional orientation is given to this 360m² space.

The designer uses a lot of smooth camber space structure, metal bead curtain, innervational camber stairs and multi-leveled green ornament to separate different functional space and keep all the areas into one core concept of the company. The exquisite plane partition meets the basic requirement to the function of a large-sized design company. The organic combination of different materials makes a blurry division of the reception room and the working place. The indeterminacy of the space makes the staff and the clients, the staff and the staff work in a relaxing atmosphere. Thus the communication between each other can be promoted and common understanding can be achieved.

在构思上设计师不是单一地从内部空间入手，而是把思维拓宽放大到有形与无形的多维空间中，从而给予了这个360平方米的空间准确的视觉与功能定位。

设计师在这里使用了大量流畅的弧形空间构成、金属珠帘、充满动感的弧形楼梯、层次多变的绿化点缀，既间隔出不同的功能空间，又使各分区统一于设计公司这个核心概念之内。巧妙精致的平面分区，满足了一个大型设计公司对功能的基本要求，而不同材料的有机结合，更使各类会客空间与办公室形成模糊分隔，这种空间的不确定性使员工与客户、员工与员工共融于轻松的空间氛围，从而促进彼此间的交流，达成一致的共识。

Shenzhen Creative-space Design & Decoration Co., Ltd. Office

深圳市创域艺术设计有限公司办公室

地　　点：深圳
面　　积：340平方米
设 计 师：殷艳明
设计单位：深圳市创域艺术设计有限公司

Walking Between the Real and the Illusive

Design is the art about space. The nature of art stems from the inner sentiments of the designer. As designer's office space, this project is not focused on piling up flaring materials or representing complex modeling, but on making full use of the language of construction to express neat scales and pleasant appearance in a disordered space, and to attain a perfect combination of function and form.

The office which has a square space is open and friendly. The three functions—reception, transportation and work—combine into a unified whole and offer the environment integrity and efficiency. Gray, the consistent color of the designing company, is chosen to inform a simple and cool-headed feeling, and various types of materials are also employed appropriately. The gray coating conveys the language of plainness and sedateness. The ash mirror displays a harmony of the real world and the illusive one. The timber bars create a compatible relationship between men and nature. All these designing elements highlight each own personality, tension and emotion, at the same time constructs an abundant sense of the vision layers and the pleasant impression of the structure.

A refined artistic character finds expression in the up-to-date screen. The reticulated woodcarving and the decorative pictures are of unique class. Time and senses have unconsciously shaped the designer's aesthetic taste. In the gray-tuned space, the warm-colored wood converses with the gray mirror, thus senses of serial and order are fully developed, and the space, people and all physical objects get rich annotations in this design.

在虚实之间徜徉

设计是空间的艺术，艺术的本质来源于设计师内心的情感。作为设计公司的办公空间，不在于花哨的材料堆砌与复杂的造型展示，如何充分利用建筑空间的形态语言，从无序的空间中通过单纯的设计语言体现有序的比例尺度和形态美感，达到功能和形式的完美结合，这才是本案追求的重点。

空间形态方正、开放而友好。接待、交通、办公功能有机结合，完整高效，在延续设计公司一贯的灰色标准色系中，各种材质的运用恰到好处：灰色涂料传达着朴实、沉稳的语言；灰镜的运用表现了虚实空间的共融；条形木材结构亲和了人的关系。各种设计元素在展现各个性、张力与情感的同时，营造了丰富的视觉层次和结构美感。

现代感十足的屏风、具有独特品味的装饰画与镂空木雕传达出脱俗的审美气质，不经意间时光与情感已历练出设计师的审美情结。灰色系空间里，暖色木与灰镜的对话，序列感、秩序感以及空间、人、物按照美学理念做出了丰富的诠释。

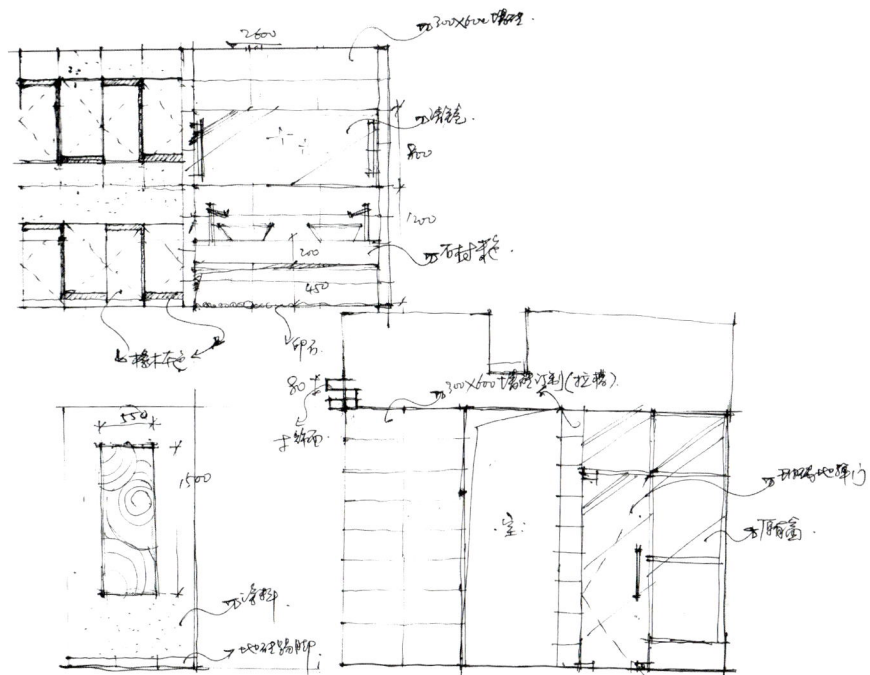

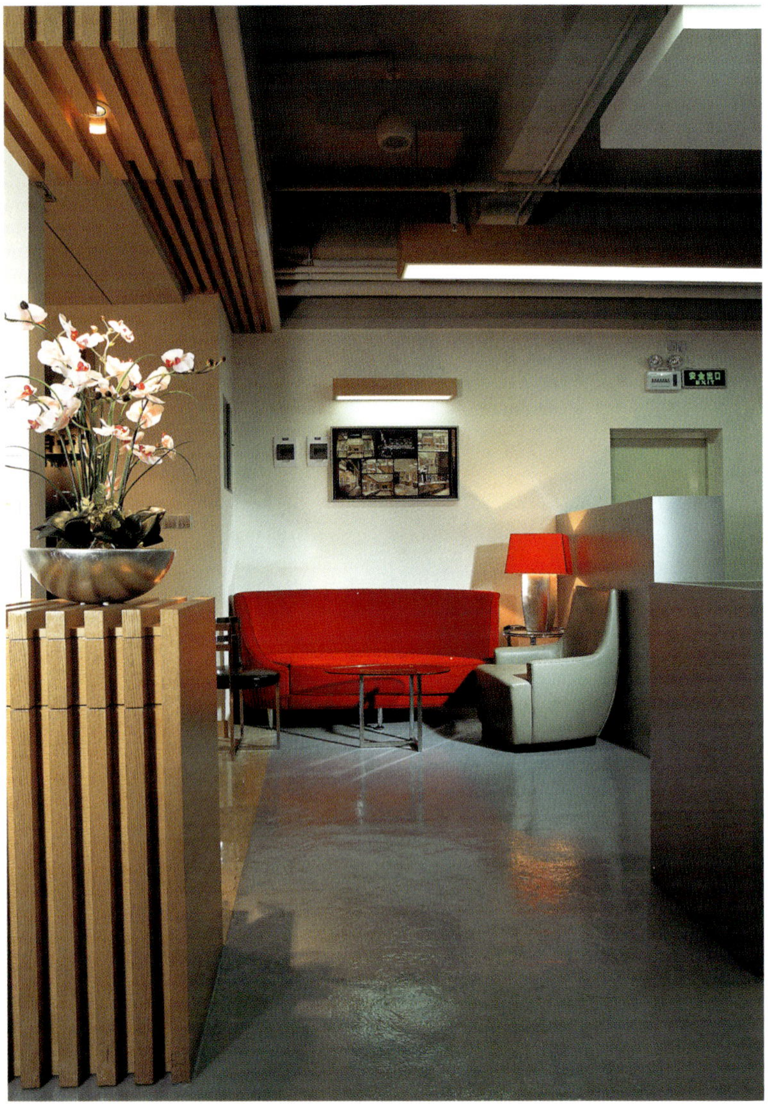

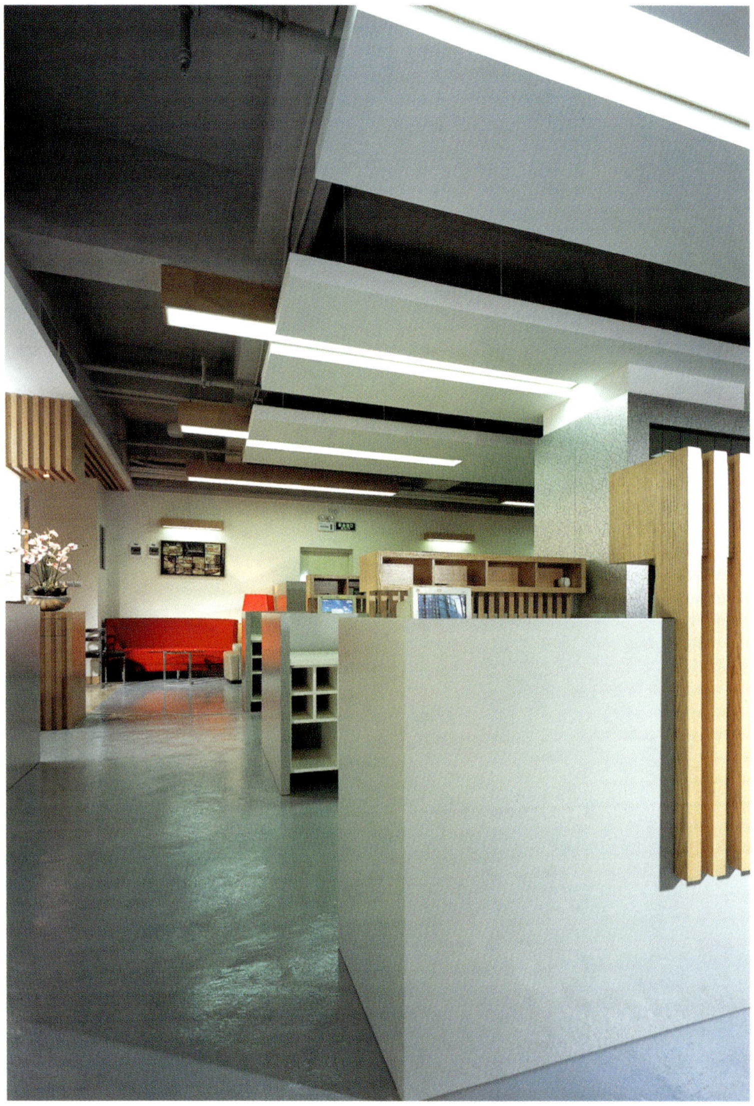

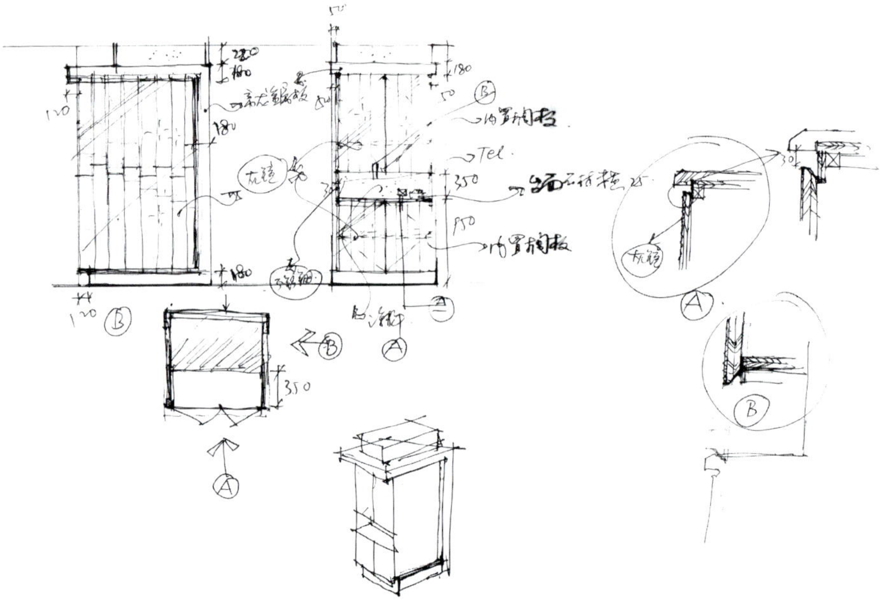

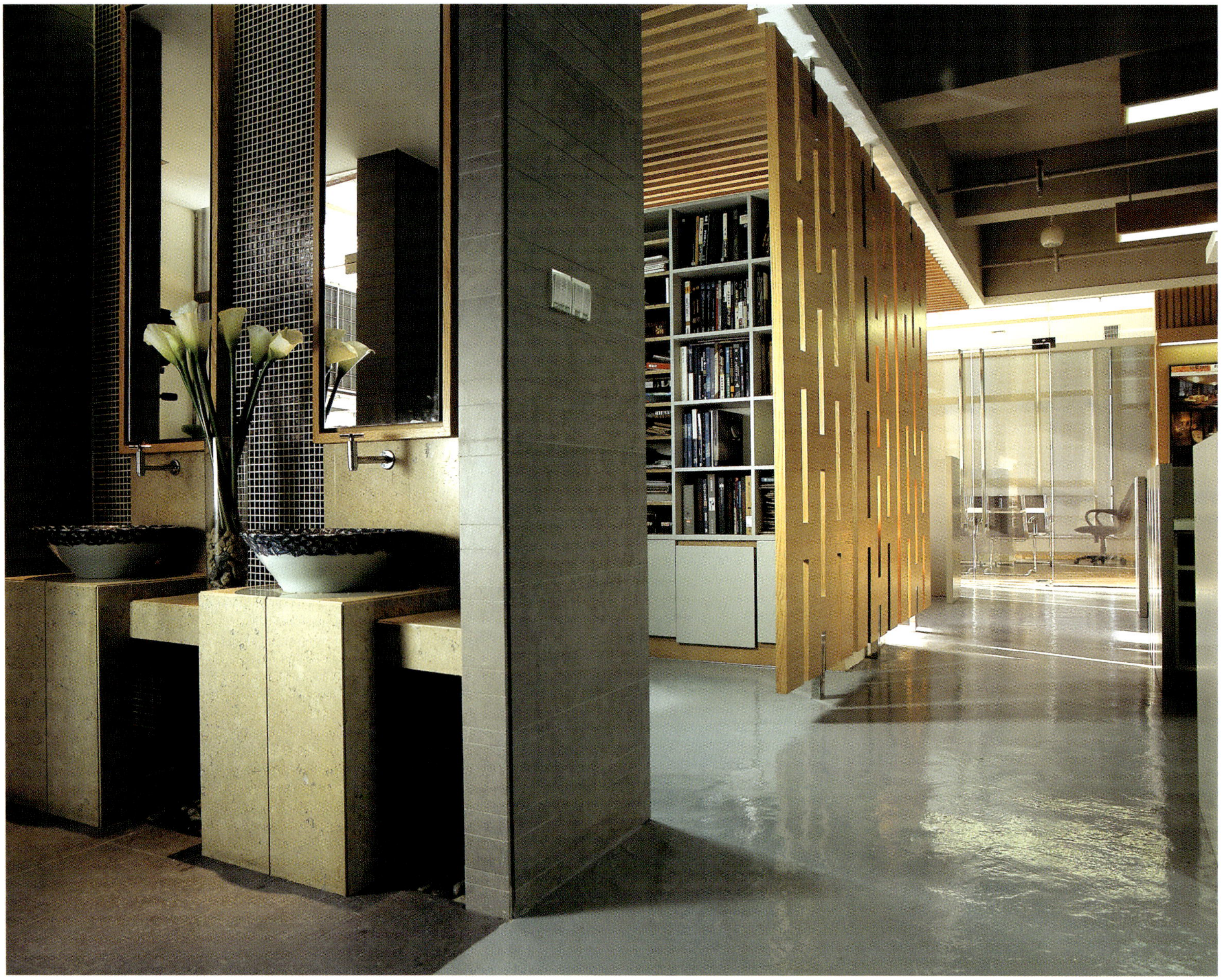

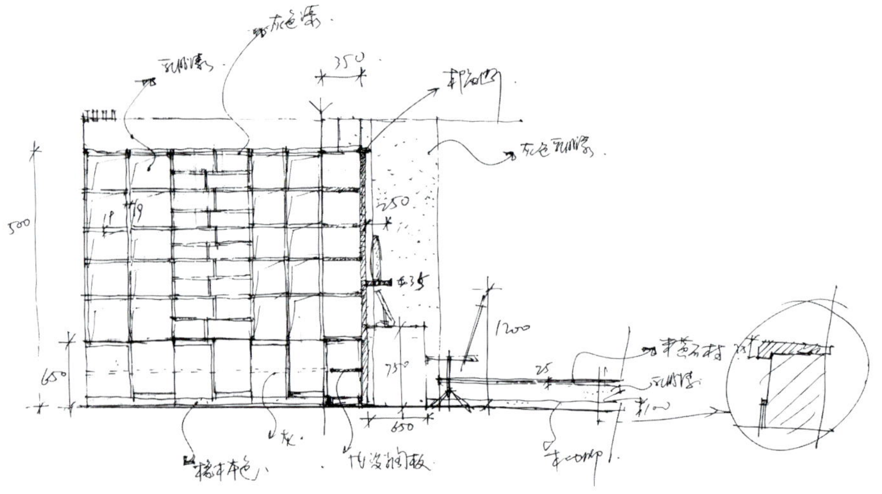

350
500
250
1200
650
750
25
650

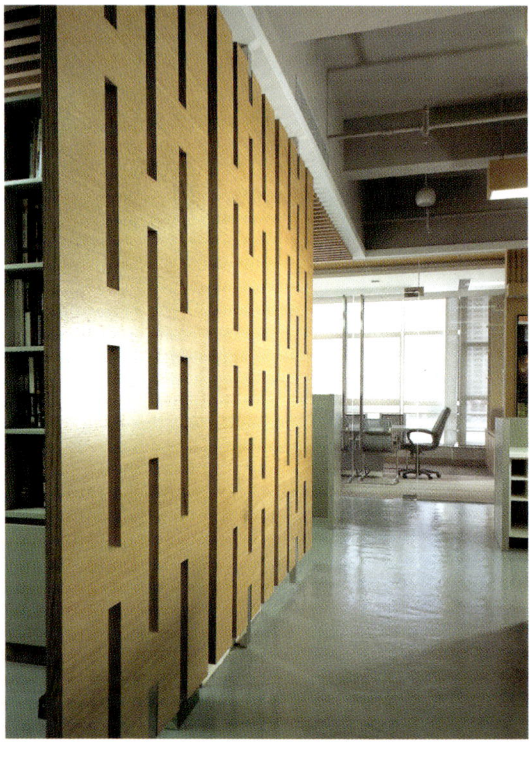

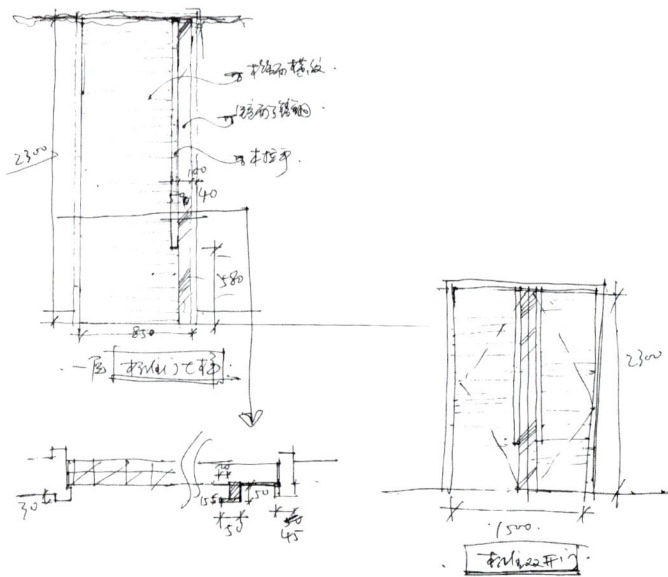

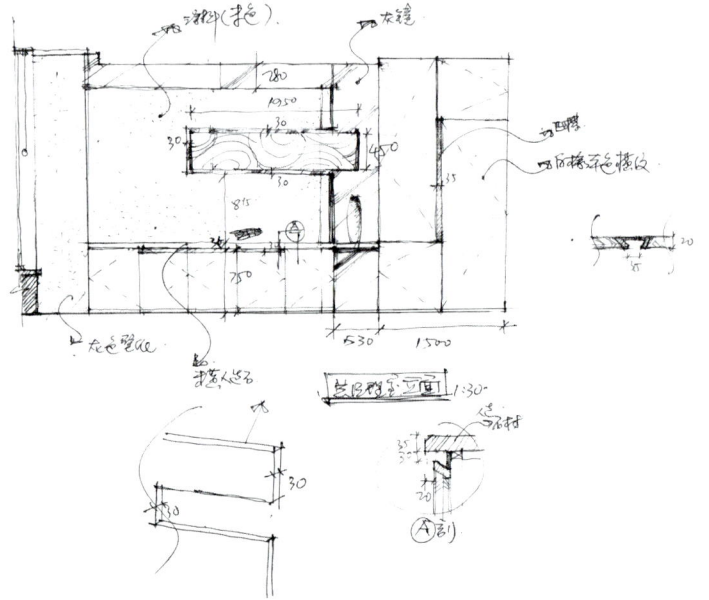

Honyoung Décor Office

泓扬装饰办公室

地　　点：深圳
面　　积：300平方米
设 计 师：杨昌龙
设计单位：泓扬装饰

The space is no longer incondite, the color is no longer depressing and the work becomes interesting. All things will come from these. The division of the bias makes the front hall into two pieces: the palavering place and the walkway. It also extends to the enlarged open working place which has the natural great momentum. The red main tinge of this building is moved to the space quietly by furniture, partition, door and window. So there is a simple and fashionable sense coming out. Using stainless steel and glass which have strong modern sense, instead of traditional wooden pattern panel, makes mutual influences with concrete floor.

让空间不再生硬，让色彩不再沉闷，让工作变成乐趣——一切皆发自于此。斜线的切割将前厅空间分成洽谈区和走道两部分，又延伸至特意放大的开放式办公区，突现大气天成之风；企业的红色主色调又被家具、隔断、门及窗台等悄然传递至空间中，无不道出简约时尚之气；抛弃传统的木纹面板，采用现代感极强的不锈钢和玻璃等材质，与水泥地面相映成趣。

C.F.L. Office in Shenzhen

C.F.L.深圳办公室

地　　点：深圳
面　　积：29000平方英尺
设 计 师：蔡明治
设计单位：蔡明治设计有限公司

C.F.L. is a Hong Kong garment manufacturer. Its Shenzhen office mainly deals with merchandising activity. Most of its clients are overseas fashion companies, the majority being casual wear. These clients would visit the office to check out the samples, so the office should look trendy, bright and welcoming.

To be in vogue, clothes could be changed frequently, while interior design should sustain constant changes. Here the designer adopts classical design element-lines, which matches the linearity of this 100-ft long office space.

Clothes are for warmth, which inspires the designer to create a warm cozy environment for the merchandisers to work in. Ash and birch are employed in this sense for wood being a warm material. These two kinds of wood share similar texture but different palette. They are put together to form some feature walls, enriching the space with their random pattern and the contrast of dark and light tones.

The low floor height and long deep space calls for a careful arrangement on the ceiling. PVC plastic tubes are hooked in neat array to create a sparse feeling as well as hiding the mechanical equipment at the same time. They are easy to dismount and install for maintenance. This also echoes with the linearity of the whole office. The false ceiling over the staff working area reflects the up lighting on top of the hanging cabinet to the desk. As a result, it avoids dazzling and also creates a solid and void contrast to the plastic tube ceiling. Several rooms have plane ceilings for sound insulation.

　　C.F.L.是一家香港服装加工公司。其位于深圳的办事处主要从事销售业务，客户多为以经营休闲服装为主的外国时装公司。由于这些客户会到办事处检查服装样品，因此这里的设计既要时尚明快又要温馨亲切。

　　为追求时尚，服装需要常换常新，而室内设计却不然，需要以不变应万变。设计师采用了经典的设计元素来配合这间100英尺的长条形办公空间。

　　服装是用于避寒的，设计师由此得以灵感，从而为经销商们营造出一种温馨舒适的氛围。采用岑木和桦木，因为木质材料会带给人一种温暖的感觉。这两种木材纹理相似，色泽却不尽相同。将它们混合使用于主题墙，其随意的图案和明暗色调的强烈对比可以很好地丰富空间。

　　由于该办事处楼层的层高不高，空间却很纵深，因此，在设计天花时需要格外用心。PVC的塑料管整整齐齐地套在一起，不仅看起来丝毫不显繁乱，同时还隐藏了机械设备，而且便于维修时安装和拆卸。此外，这一设计与整个办事处的线形格局也相得益彰。员工工作区上面的假平顶将吊橱上方的照明灯光反射到桌子上，避免了令人目眩的强光，并与铺设着塑料管的天花形成了虚实的对比。为达到隔音的效果，几间房间的天花采用的只是光滑的平面。

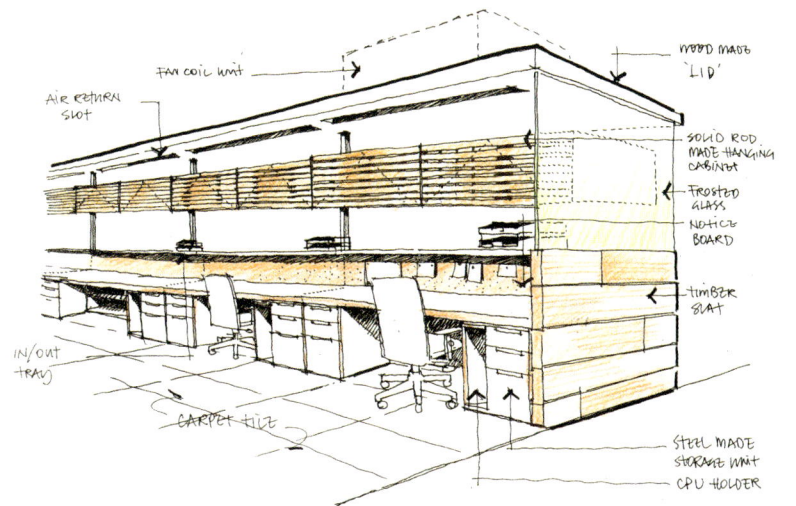

Glorious Sun Enterprises Ltd. Headquarters

旭日集团总办事处

地　　点：惠州
面　　积：13000平方英尺（2层），18000平方英尺（3层），36000平方英尺（10、11层）
设 计 师：蔡明治
设计单位：蔡明治设计有限公司

Headquarters of Glorious Sun Enterprises Ltd. in Huizhou is on the second and the third floor, where is located with the group president's office room, general manager's office room and business conference room. The design is made after considering president and general manager's basic requirements. The style and colors chosen are sedate and explicit. On the basis of simplicity, practical use and neatness, much consideration is taken into details; therefore, they present the sedateness and individuality at the same time.

On the 10th and 11th floor of the building is Jeanswest, the retailing garment brand of the Glorious Sun Enterprises Ltd.. This brand expresses a feeling of youth, comfort, unconstraint and vigor. As a result, to match with this brand's image, an irregular technique is applied to deal with the lighting and positions of the airconditioners on the ceiling, and the monotone white is set off by colored lines. Individuality and originality are displayed through the plane design in the room and direction plates, making the whole office more vigorous and out of ordinary.

旭日集团位于中国惠州的总办事处位于2层及3层，主要是集团总裁及总经理的办公室和商务会议室。设计方面，先了解总经理及总裁的基本要求后再考虑设计。设计及颜色的运用较为稳重、明确，以简单、实用及整洁为基本原则，再在细部上花心思，表现集团稳重的同时亦有风格。

旭日集团旗下的零售服装品牌"真维斯"位于大楼的10层及11层。"真维斯"品牌给人的感觉是年轻、舒服、无拘无束和充满活力。所以在设计方面亦配合其品牌的形象，以不规则的手法处理天花的灯光及冷气风位，再利用颜色的线条突出单调的白色，最后运用平面设计在房间及指示牌上突显个性及创意，为整个办公室环境增添不少活力及个性。

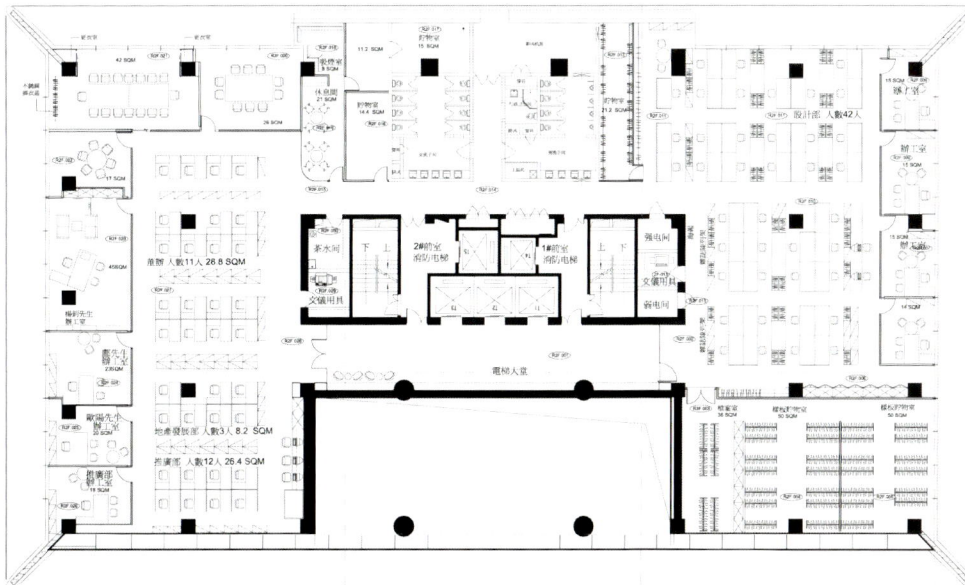

smoking room
吸烟室

I-one Financial Press Ltd. Office

卓智财经印刷有限公司办公室

地　　点：香港
面　　积：870平方米
设 计 师：蔡明治
设计单位：蔡明治设计有限公司

Located in Central, Hong Kong, i-one Financial Press Ltd. specializes in financial press, whose clients are large corporations and transnational investment banks both in Hong Kong and abroad. The designer plans his design on the basis of the quality of the clients' business. Therefore, two main areas are partitioned in the plan: the area exclusively for staffs and the area exclusively for customers. The area exclusively for staffs, including the office area, manager office and meeting room, is designed with the principle of simplicity. Whereas the focus of the whole design is the big rest area and the big and small meeting rooms in the reception center and the area exclusively for customers. The main principle of the design is briefness, novelty and comfort, which provides clients a comfortable place to linger on and to work.

位于香港中环的卓智财经印刷有限公司专营财经印刷业务，针对的客户是香港及海外大型企业和跨国投资银行。设计师根据客户的业务性质而设计，所以在平面设计上划分了两个主要区域："员工自用区"和"客户专用区"。"员工自用区"以简单为设计原则，包括办公区，经理室及会议区。而接待处及"客户专用区"的大型休息区及大小型会议室是整个设计的重点所在。设计以简洁、新颖、舒服为主要原则，令客户有一个舒服的地方长时间逗留和工作。

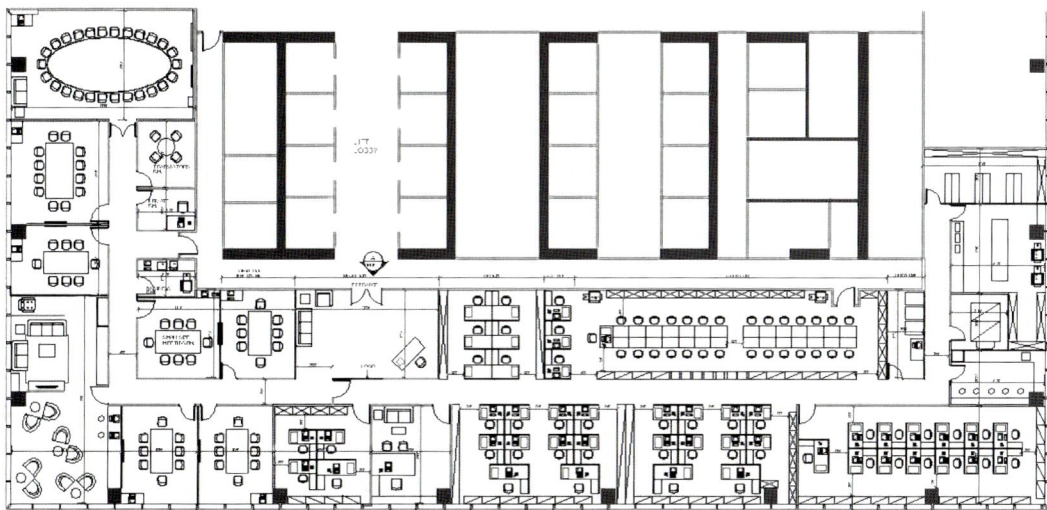

Standard Chartered Bank (Hong Kong) Finance Studies Center

渣打银行(香港)金融研习中心

地　　点：香港
面　　积：2860平方米
设 计 师：邓子豪
设计单位：天豪设计

Occupying two joint floors, this institute is located in a building on Causeway Bay, Hong Kong. Under the frame of the practical construction, a strong visual impact is created by particular materials and the extension of bright colors. A dynamic repetition of color is applied in these two floors. On entering the reception hall, students will be attracted by the special wall. Proper arrangements transfer the plane elements to three-dimensional sights. Waves of ripples cover the whole wall, which is not only a part of imago, but also is spread out on purpose. Moreover, out of the waves a niche is carved, which is made as a small aquarium, adding more vigor to the environment.

Varied colors convert dull classrooms into cozy learning environment. These classrooms differ in colors and consequently and bring much vigor to the corridor. To make the environment more lighthearted, classrooms are named after different countries instead of the conventional ways.

　　占用了连续两层的学院位于香港铜锣湾区的一栋建筑内。在实用的建筑外框下，透过独特的材料与鲜明色彩的延伸，造成视觉上强烈的冲击。动感的色彩重复使用在这两层楼上。学员一进入接待大厅就会被有特色的墙壁所吸引。恰当的安排把平面元素转化为三维的空间景观。层层的波纹覆盖整个墙壁，其不只是意象的一部分，还是有意的铺排。此外，波纹外面还雕刻了壁龛，并设置了小型水族箱，营造出更加充满活力的气氛。

　　乏味的教室因多种色彩的使用而变成了温馨的学习乐园。教室的颜色各不相同，造就了一个充满活力的走廊。为了增加轻松的气氛，教室选用不同国家的名字来命名，以取代传统教室的称呼。

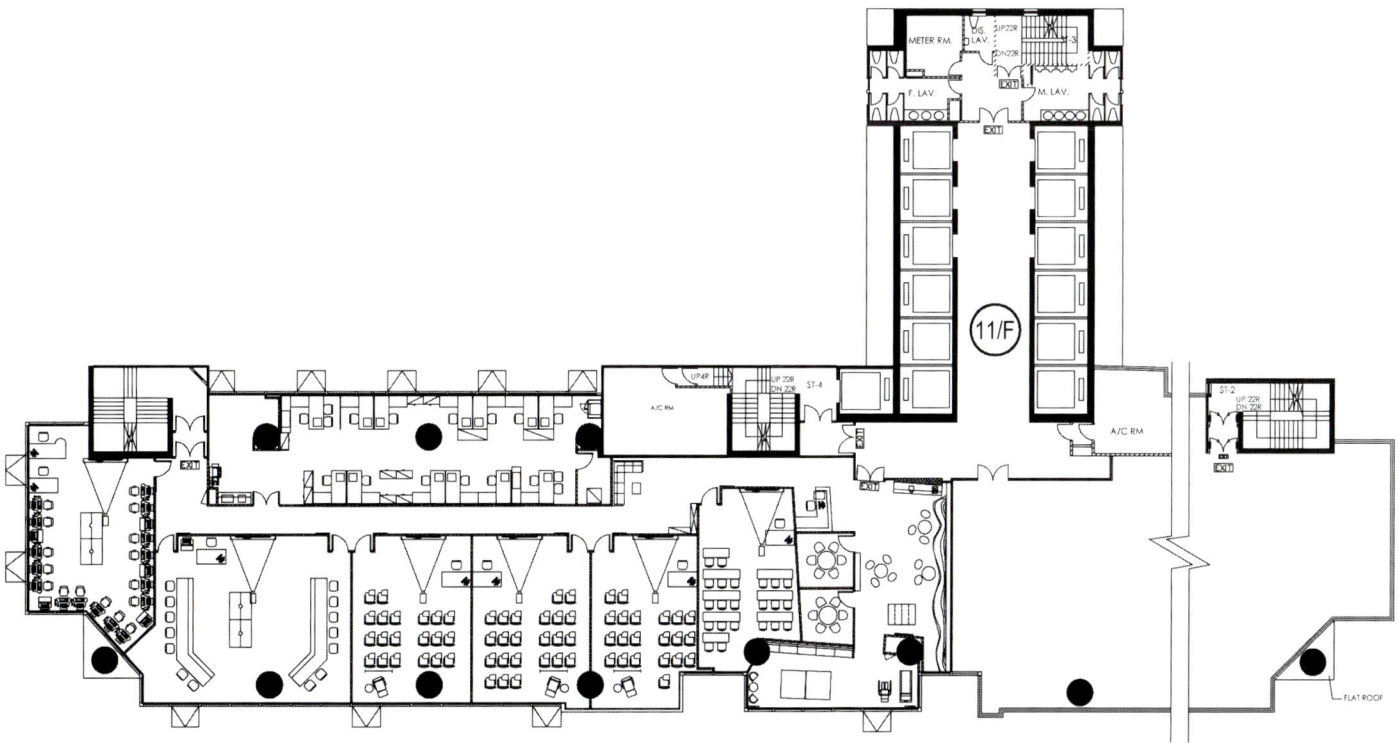

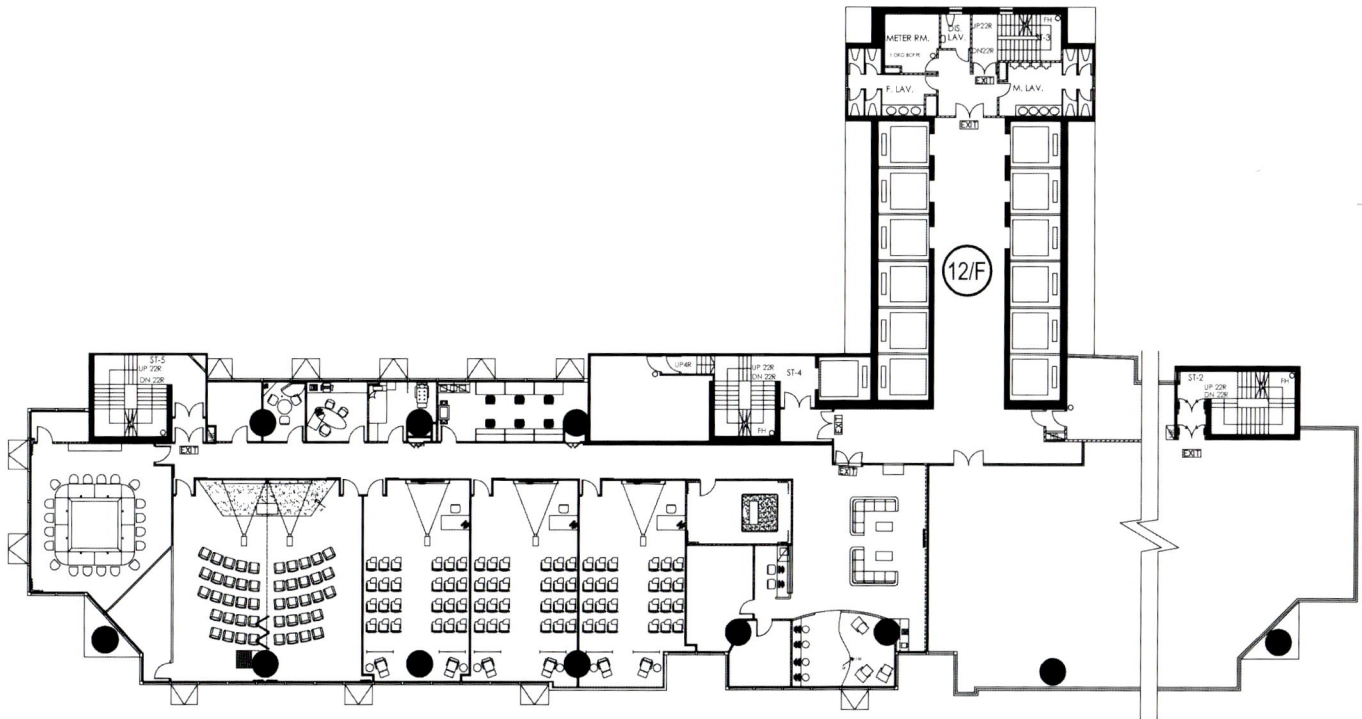

India

Goldstar Communication Technology Co., Ltd.

高盛达通信技术有限公司

地　　点：深圳
面　　积：1300平方米
设 计 师：陈飞杰
设计单位：深圳汇杰室内设计工程有限公司

As a communication technology company, the designer uses very simple design ideas to give this 1,300m² office a new breath. He uses the contrast between natural marble wall and bright artificial stone floor at the entrance of the front hall to reflect a different working environment from others by special lighting effects.
All the Air Conditioners and extinguishers are all bare outward under the ceiling at the large working place which has a very dense industry taste.
The space inside is very fluent, and the mixture of color is simple. Vogue and nature supplement each other. The designer uses his own design language to create a flexible and smart space.

作为通信技术类公司，设计师用非常简洁的设计语言给予了这个1300平方米的办公室新的气息。大前厅入口处采用天然大理石墙面与亮光人造石地面做对比，以特殊的光照效果烘托出了其与众不同的办公环境。

大办公区天花将所有空调、消防等系统裸露在外，工业味道浓厚。

室内空间流畅，色彩搭配简洁，前卫与自然相辅相成。设计师用自己特有的设计语言营造出一个灵动的空间。

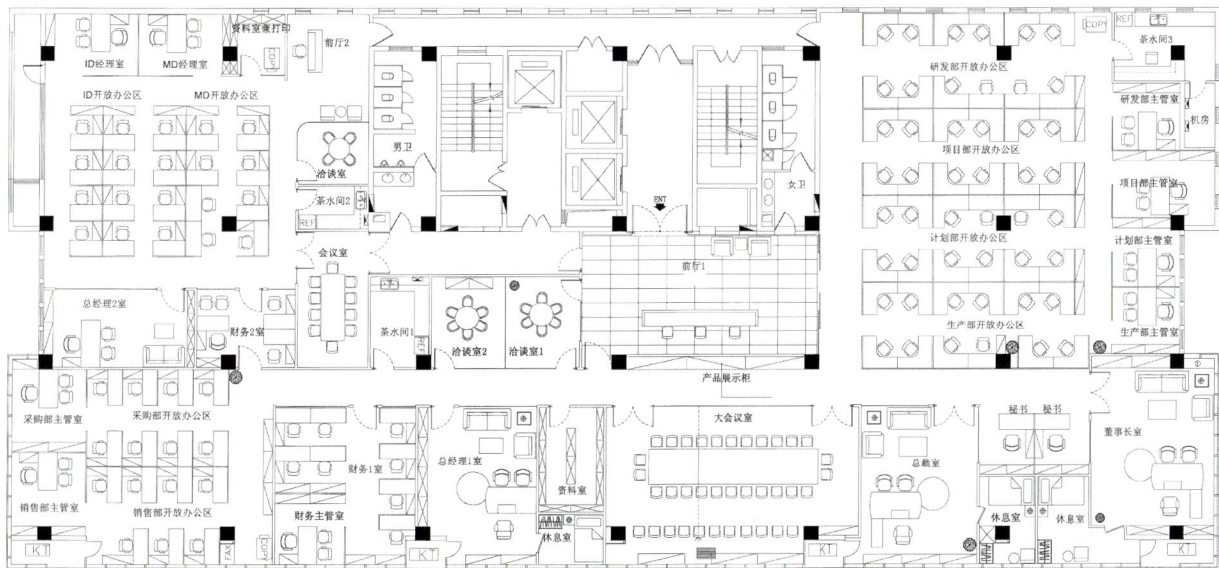

Wen Hua Tenement Office

文华地产办公室

地　　点：西安
面　　积：1000平方米
设 计 师：龚小刚
设计单位：龚小刚设计师事务所

This company is a company which generalized small size apartments as a beginning. This office is on the top floor of Xi'an Wen Hua Building. After 10 years' development, the company decides to do decoration and modification to the old building.

Financial department and cost department, having functions of external liaison and reception, are on the left side of the hall. They are close to the negotiating area、toilet and the water bar. So it has the dynamic property. The administration department, next to the general manager's and the vice manager's office, is an internal communicating room which is half-dynamic. The Director's office is at the end, which contains movie room、negotiating room and the green audience meeting room. By using the symbol of fuzzy concept all way long, properties of multi-elements、multi-culture and multi-behavior combined together and putting firmness through effeminacy of the female leader can be expressed. In light of the exact orientation of the above three spaces, the space is thus visualized appropriately and accurately.

The offices use extremely modern and neoclassic decoration which grasps features of fortune, character, fad and culture well and truly.

该公司是一家以主推小户型住宅起家的地产公司，办公室位于西安文华大厦顶层。公司经历了十年的成长历程后，决定利用旧建筑进行装修改造。

财务部、造价部是对外联络和接待的空间，布置在门厅的左侧，紧邻洽谈区、卫生间、水吧，具有动态的属性。行政等部门是内部的沟通空间，紧邻总经理室和副总经理室，具有半动态的属性，董事长室位于最尾端，配置了影视厅、谈判间、绿色接见厅，利用一个概念模糊的符号贯穿始终，具有女性董事长多种元素、多种文化、多种行为兼收并蓄、柔中有刚的空间属性。鉴于对上述三个空间的准确定位，空间形象的产生就十分贴切、准确。

办公室采用极端现代和新古典主义的配饰，准确地把握了财富、品格、时尚、文化的特征。

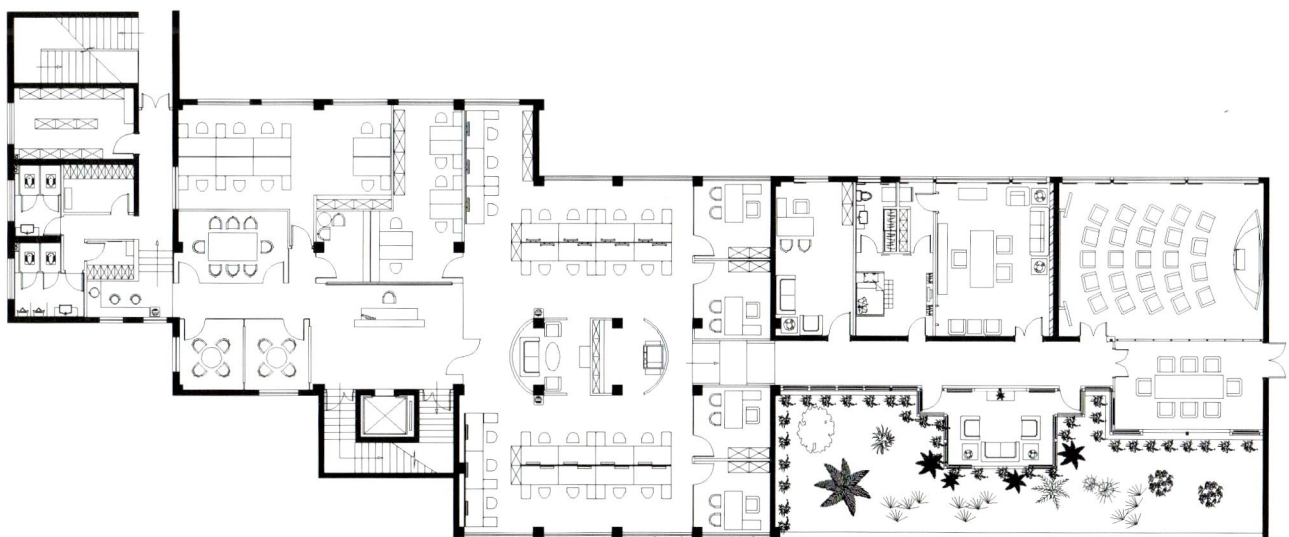

An Office in Wenzhou

温州市某办公室

地　　点：温州
面　　积：140平方米
设 计 师：曾建龙

The room design of this case is different from traditional ones. It is expressed by the form of chamber in the whole concept. Although the space is not very large, the designer adopts the form of half-open in the room structure. The combination of glass and wood ornaments, the art treatment of marble, the use of brown mirror and the pulse use of carpets all pivot on the consistency of the room and express this subject. The balcony resting place is designed to create an atmosphere of southeast style, where you can read, drink tea and chat. You can feel the frame of mind with ease and harmony. And the humanization of the new office room can be transmitted.

本案办公空间区别于传统的办公空间设计，在整体的概念中以会所的形式去表现。虽然空间并不是很大，但是在空间结构层次上的运用却以半开放形式体现，玻璃和木饰面的结合运用、大理石的艺术处理、茶叶色镜面的运用以及地毯的动向运用都围绕和体现空间的连贯性。阳台休闲区被设计成带有东南亚风格的氛围，在这里可以看书、喝茶、聊天，可以去感受一份轻松和谐的心境，从而传达了新办公空间的人性化。

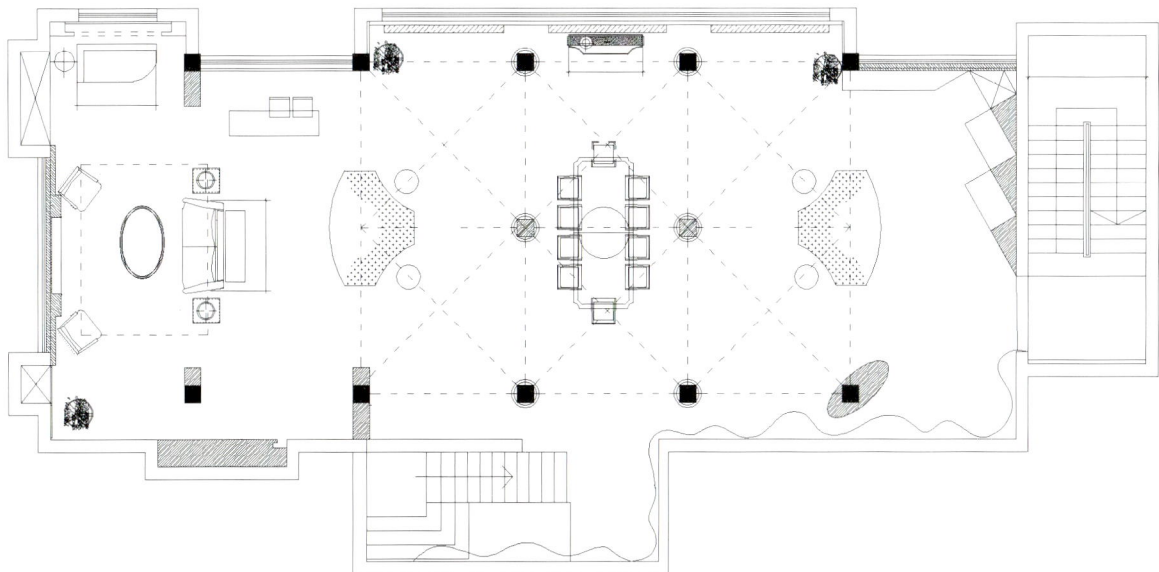

international
council of interior
architects
& designers

G &

SINGA
INTER
ARC
DESIG

PD

PHITA
DESIG
INDUS
SDN
BHD

China Marketing

《销售与市场》杂志社

地　　点：郑州
面　　积：1450平方米
设 计 师：陈思宇
设计单位：河南贝铭设计装饰工程有限公司

When standing in the space defined as the round sky and round earth, in front of the door-like-pattern construction of the reception desk, people feel that they are standing in the restriction of the space. Through the different sizes but the same shape of the window hollow, the newly-opened waiting area makes the atmosphere become relaxing but adding tones of mystery.

The whole office adopts only yellow and brown to decorate. The floor adopts inoculation block. The decoration of the wall adopts changeable colour in terms of construction style; the mini-sized green plant is growing on the short partition wall. There are also scattered potted plants like fernleaf hedge bamboo, Pachira macrocarpa, which have entitled busy people a kind of pastoral feeling of relaxation, nature and refreshment.

In some cases of space's partition treatment, the designer adopts thin transparent soft yarn to decorate, which makes the officers feel at home, and represents the compatible side of modern management.

在有如门式图案组合的接待台前，人们站在天圆地圆的空间限定中。旁边开辟出的休息与等待的小场地，透过镂空的大小各异的又有统一形状的窗洞孔，使轻松中又带有几分神秘。

整个办公场所，只采用黄色与褐色，地板采用拼接木块，墙面砖使用略有色差的砌体。矮隔断上种植的绿意浓浓的微型植物，还有散置的凤尾竹、发财树等盆栽，给紧张忙碌的人们注入了田园般轻松、自然、清新的感受。

在有些空间的隔断处理上，设计师采用了薄薄的透明轻纱，也给了办公人员以家的感觉，体现出现代管理亲和的一面。

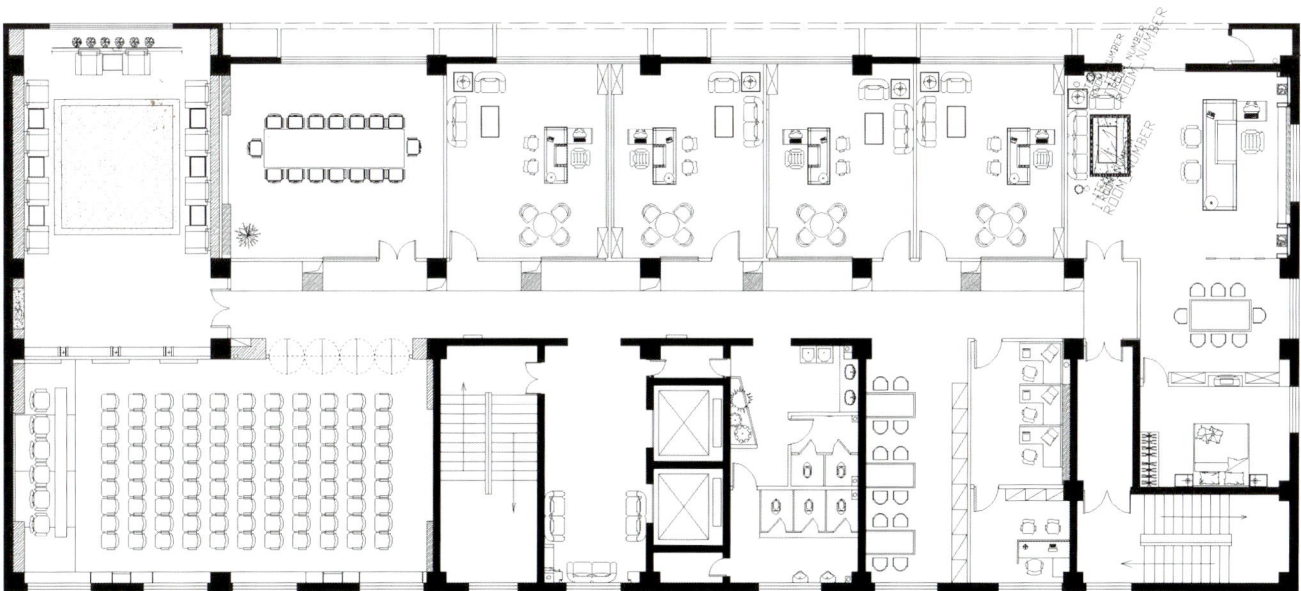

IN · X Designing Office

IN · X设计工作室

地　　点：北京
面　　积：132平方米
设 计 师：吴为
设计单位：北京（IN·X）屋里门外设计公司

The designer chooses gray as the main tinge of the space specially, the purpose of which is to meet the functional requirement of the thinking workshop. The red in it mobilizes the vital force in the gray tinge. The combination of red and black is always giving a classical appearance. The most conspicuous red appears on the balustrade of the stair to the second floor. The obvious red metal welted line adjusts the visual singularity of the large part of gray, and also has some warning effects to this area.

The working area on the first floor can be shared by 6 or 7 people. In the center of the working place are the coping desk and paper cabinet. When communication and discussions are needed, the designers can complete their projects right here, which is very convenient and very flexible. The palavering place to receive guests is located in front of the open style kitchen which is used as a water bar. Here they can have small meetings and discuss schemes. As the raising up of the working place, there is a division between the reception and working place, where there will be no interruption between these two areas. The wall which face the window is taken away in the open study on the second floor. So the space, open and running through, is called by designers "the place to have thought communication with masters".

设计师特意挑选出灰色作为空间的主调，为的便是满足"思想车间"的功能需要。跳跃其中的红色将灰调子中的生机调动起来，红与灰黑的搭配永远是以经典的身姿出现。空间中最为出彩的红色出现在通往二层的楼梯栏杆上。醒目的大红色金属折线调节了大面积灰色所带来的视觉上的单一，同时也在这一区域起到了警示的作用。

一层的工作区，可以供6、7人使用。工作区的中间位置安放了拷贝台与图纸柜。当需要交流与讨论时，设计师们可以很方便地在此完成工作，使用上非常灵活。接待客人的洽谈区设在作为水吧使用的开放式厨房前，在这里可进行小型会议或是探讨方案；由于工作区部分的地面被抬高，接待区与工作区在空间上有了划分，使用中便可互不干扰。二层的开放书房面向窗子的墙面被完全拆除，将空间打开，上下贯通。这里被设计师命名为"与大师思想交流的地方"。

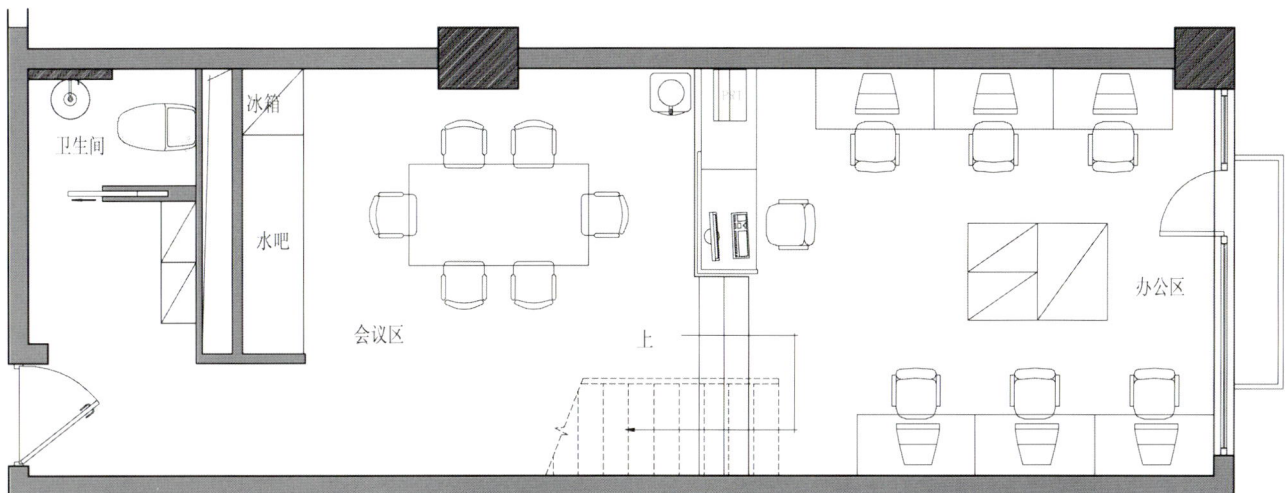

卫生间

冰箱

水吧

会议区

上

办公区

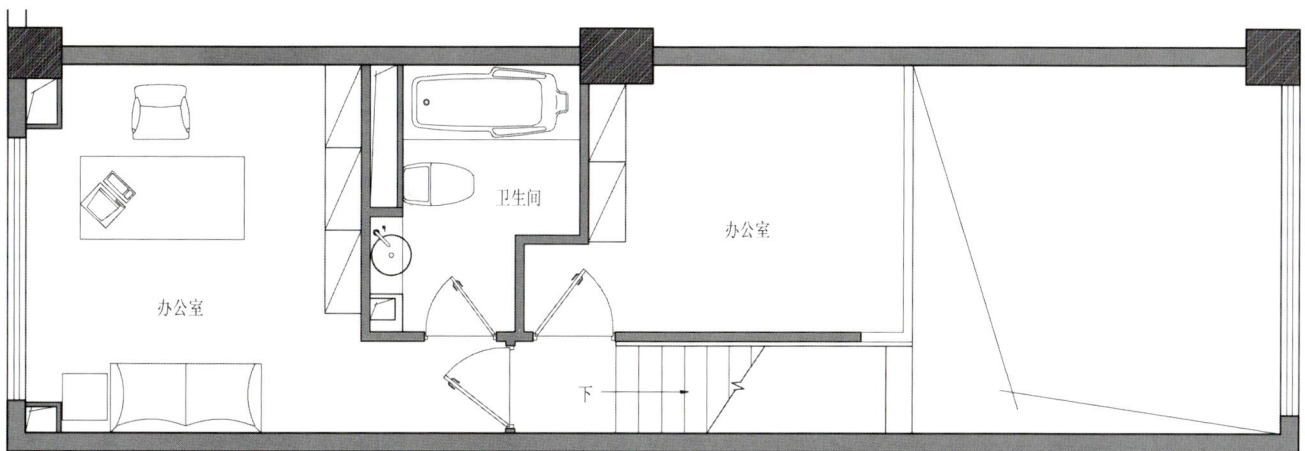

办公室

卫生间

办公室

下

City Building

城市大厦

地　　点：深圳
设 计 师：冯烈　焦山
设计单位：深圳市美佳装饰设计工程有限公司

Designers bring in an office design concept which is healthy and humanized at first. They use bright colors, simple mould and environmentally protected decoration materials to create an inner space which is full of vitality, human cultural sights and high balance. It shows the modern charm which the "after-SARS period" has.

　　设计师首先提出了一个健康的和人性化的办公空间设计理念，采用明快的色调、简洁的造型和环保的装饰材料来营造一个充满生机、人文景观和高度平衡的室内空间，展示出它"后SARS时代"所具有的现代魅力。

Ai Bao Technology (Hong Kong) Co., Ltd.

爱宝科技（香港）有限公司

地　　点：香港
面　　积：350平方米
设 计 师：陈飞杰
设计单位：深圳汇杰室内设计工程有限公司

Ai Bao Technology (Hong Kong) Co., Ltd., a mature technology limited company, deals with video and multimedia packages. It strives for technological innovation, excellent quality, human-orientation and the harmony of nature and technology. For this reason, throughout the plan, the designer follows the principle of fully combining the practical and functional requirements with humane management, which links office needs with the working flow, and partitions different functional areas in a reasonable way to meet the requirements of office business. In addition, they attach much importance to the simplicity, beauty and good taste, presenting the company's profile and expressing a modern sense. In the design of the space, basing on the style of briefness and modernness, the designer places emphasis on the human design concept of practicality, modernness and environmental protection. In the color of the space, gray-the company's imagery color-is employed to enforce the company's strategy of CI, and to depict the Hi-tech image of the company. Self-leveling materials are employed on the floor, and this kind of materials is of wear-resistance and good taste, and also easy to clean. On top of that, it looks noble, elegant and grand, which gives expression to the company's capability and image. Matching with the style of the floor, the bare ceiling and gray tone look brief and clear-sighted. With the reasonable color match and soft lighting from the tube-shaped lamps, the whole lobby demonstrates a cultural environment of a modern corporation's office.

爱宝科技（香港）有限公司是一家成熟的经营音像及多媒体包装的科技有限公司。公司力求科技领先，追求卓越品质，倡导以人为本、自然和科技的和谐统一。因此，设计师在平面规划中自始至终遵循实用、功能需求和人性化管理充分结合的原则，既结合办公需求和工作流程，科学合理地划分职能区域，以满足办公的需要，又注重形式上的简洁、大方、美观，充分体现出企业的形象与现代感。

在空间的处理上，设计师以简约、现代的设计手法为依据，重点致力于体现实用、现代、环保的人性化设计理念。在空间色彩上，运用该公司的企业形象色——灰色，以配合企业的CI策略，体现高科技企业的形象。地面采用自流平材料，具有大气、耐磨、易清洁的特性，同时还能给人高贵、典雅、庄重的感觉，能体现出公司的实力与形象。天花裸露，喷灰色调，和地面风格统一，简洁而明快。合理的色彩搭配及筒灯柔和的灯光设计，使整个大堂呈现出现代企业办公环境的文化氛围。

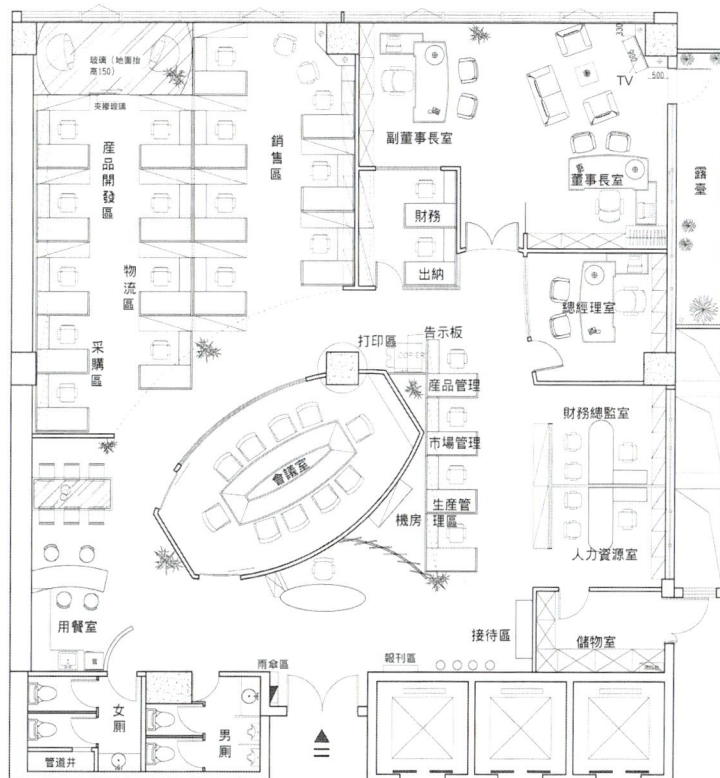

玻璃(地面抬高150)

夾層玻璃

產品開發區

物流區

采購區

用餐室

管道井

女廁

男廁

雨傘區

銷售區

副董事長室

財務

出納

董事長室

露臺

總經理室

打印區

告示板

COPER

產品管理

市場管理

會議室

生產管理區

機房

財務總監室

人力資源室

接待區

儲物室

報刊區

Ante Medical Company Office

安特医药公司办公室

地　　点：汕头
面　　积：370平方米
设 计 师：陈骏 林影
设计单位：汕头市蓝鲸室内设计有限公司

On the design of sculpt, the designers use transparency as the main method: to use large area of floor glass curtain wall and steel structure as area division; to bare the ceiling structure to get the transparent feeling; to put gray and blue mirror to the central lift room which makes a mirror block. This blurs the ponderosity of the lift room and spreads the moving line of the whole space. By using the refraction of mirror medium like magic block, the space is extended and unfolded from visual sight which makes it full of expansibility and continuity. And this makes the space connection and communication possible.

In the aspect of sculpt method, designers use penetrable wall and blocks as the sculpt elements of the moving design. These transparent walls and blocks of different dimensions and scales are arranged orderly, and an ordered partition of the whole space is thus created by the directional misalignment, complementation and connecting of them. Apart from that, they are also endowed with a strong function of guiding the space.

In the aspect of color, the design uses gray blue as the main color. Some local warm colors are used on the lift door, the entrance hall, the meeting room, the terrace and the end of the walkway and so on where people will stop. While reducing the narrow and depression of the walkway, the refraction of the mirror is used to achieve the moving of colors and the mutual penetration of the space.

　　在造型设计上，设计师以"透"为主要手法，运用大面积的点式落地清玻幕墙及钢构件进行区域划分，裸露天花结构以获得通透的场感；将位于中央的电梯间贴上灰蓝镜面，使其构成一个镜面方体，不仅模糊了电梯间的笨重，而且由此展开整个空间动线；犹如魔方一样，借助镜面介质的折射，从视觉角度将空间延伸、展开，使之更富有扩展性、延续性，使空间的连接及对话成为可能。

　　在造型手段上，以"可以穿透的墙、块体"作为空间动线设计上的造型元素，将不同体量、比例的"透墙、透块"通过有序的设置及方向的错位、互补、衔接，对整个空间进行有序的分割，同时赋予强烈的空间引导作用。

　　在色彩方面，以灰蓝色调为主，将局部暖色运用于电梯口、门厅、会议室、露台、通道尽头等动线人流的驻足点，在淡化通道狭长、沉闷的同时，通过镜面的折射获得色彩的流动及空间的互相渗透。

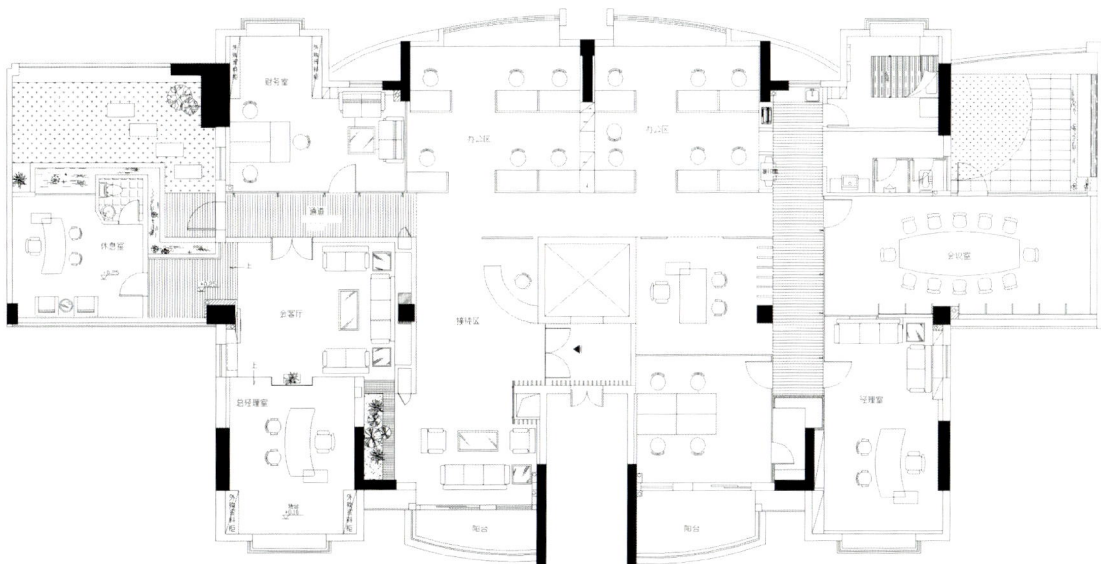

Huzhou Kaier Fashion Office Building

湖州凯尔服饰办公大楼

地　　点：湖州
面　　积：3000平方米
设 计 师：陈品浩
设计单位：宁波柏天设计公司

Conventional and unconventional

This project is a rectanglar space with the atrium as the center. In order to neglect the conventional inflexible space layout, the designer made "unconventional" elements applied in the "conventional" space, making use of ruleless shape to neglect rectanglar layout to make the space become lively. The reception space's ruleless geometrical shape formed amusing shadow varying, producing forceful conflict and contrast on vision, and achieved harmony of modern and simple through the choice of white.

常态与非常态

　　本案是一个以中庭为中心的矩形空间。

　　为了打破呆板的常态空间布局，设计师以＂非常态＂元素在＂常态＂空间里的变化运用，借无规律的形态打破矩形的布局，使空间变得生动；接待区无规律的几何形态形成有趣的光影变化，在视觉上产生强烈的冲突与对比，又在白色的纯净基调中实现了现代与质朴的和谐统一。

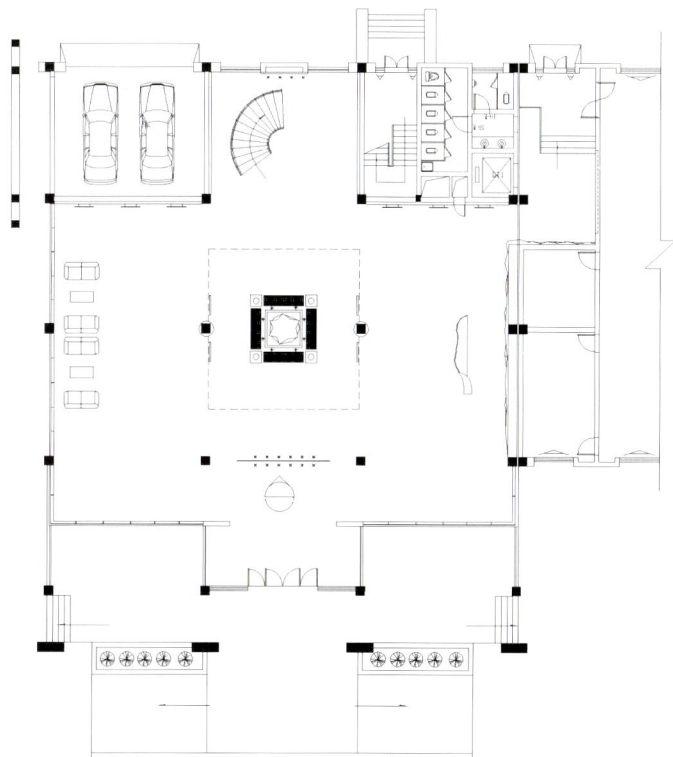

Howyawl Group Office Space

浩源集团办公空间

地　　点：青岛
面　　积：600平方米
设 计 师：李明
设计单位：Free3man设计顾问公司

The office space was originally the enterprise's dining hall. Through the alteration of the old space, the designer integrated the complicated structural style with the approach of getting through the space, to strengthen the function of the whole space. The entire style is modern, concise, and humanistic. Alteration of space as the clue. It maks use of the materials, such as steel, bamboo, glass, to control the cost. Using grey, blue and white as the main color to emphasize the character of the enterprise. Haoyuan Group manifests enterprise's vigor through the connotation of its space.

　　办公空间原为企业的食堂。设计师通过对旧空间的改造，将复杂的结构形态均打通整合，使整体空间功能性增强。整体语言为现代、简约、人文形态。以空间改造为线索，运用钢、竹、玻璃这些材料，控制造价。颜色为灰、蓝、白，突出企业性质。浩源集团以空间内涵彰显了企业的活力。

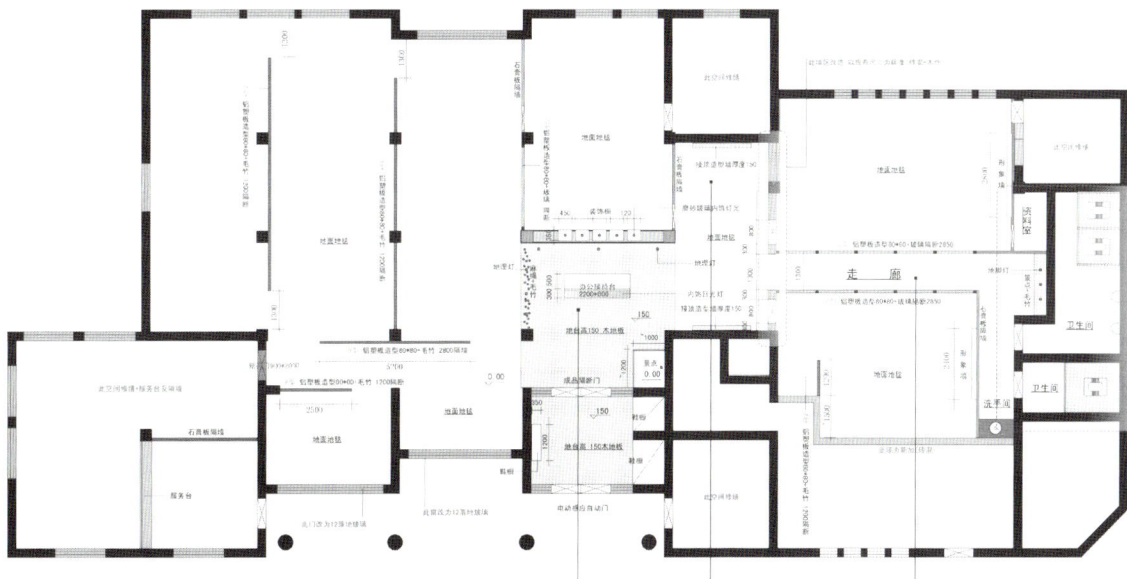

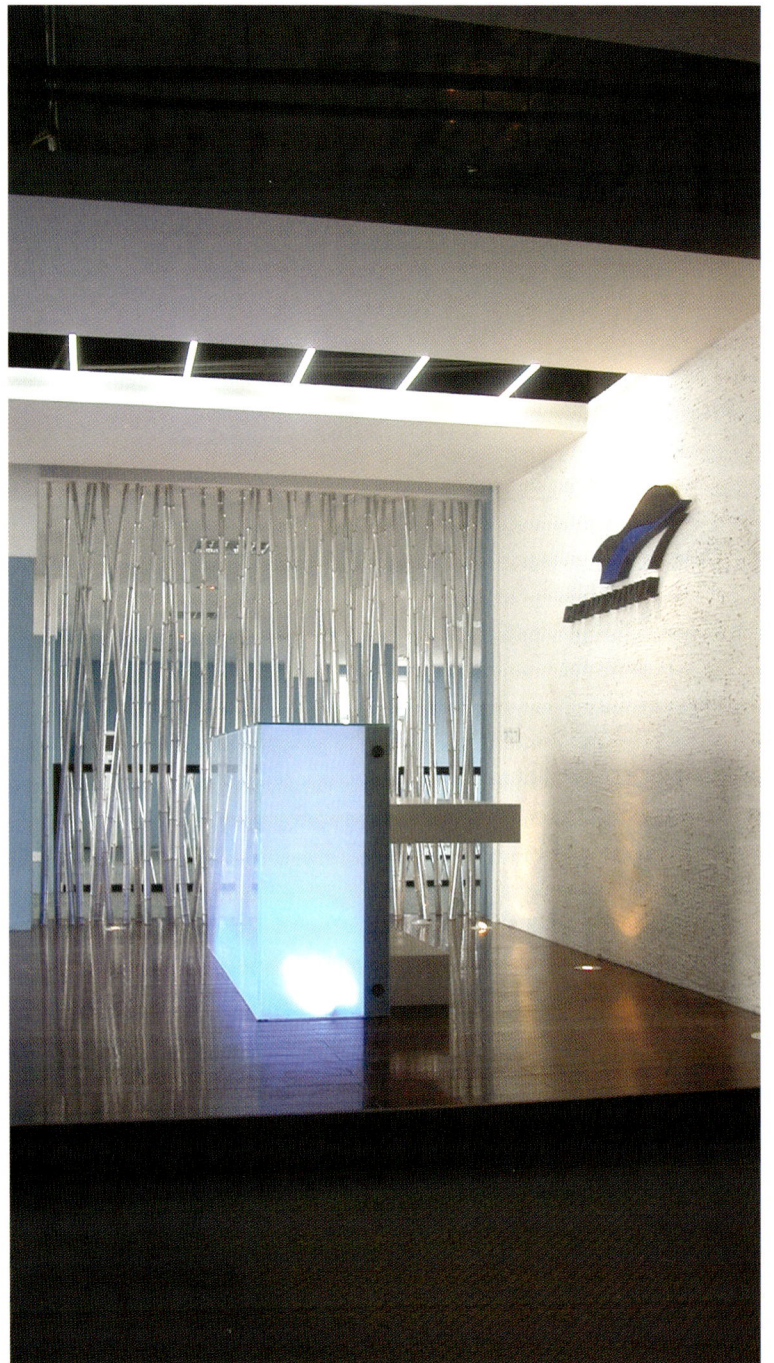

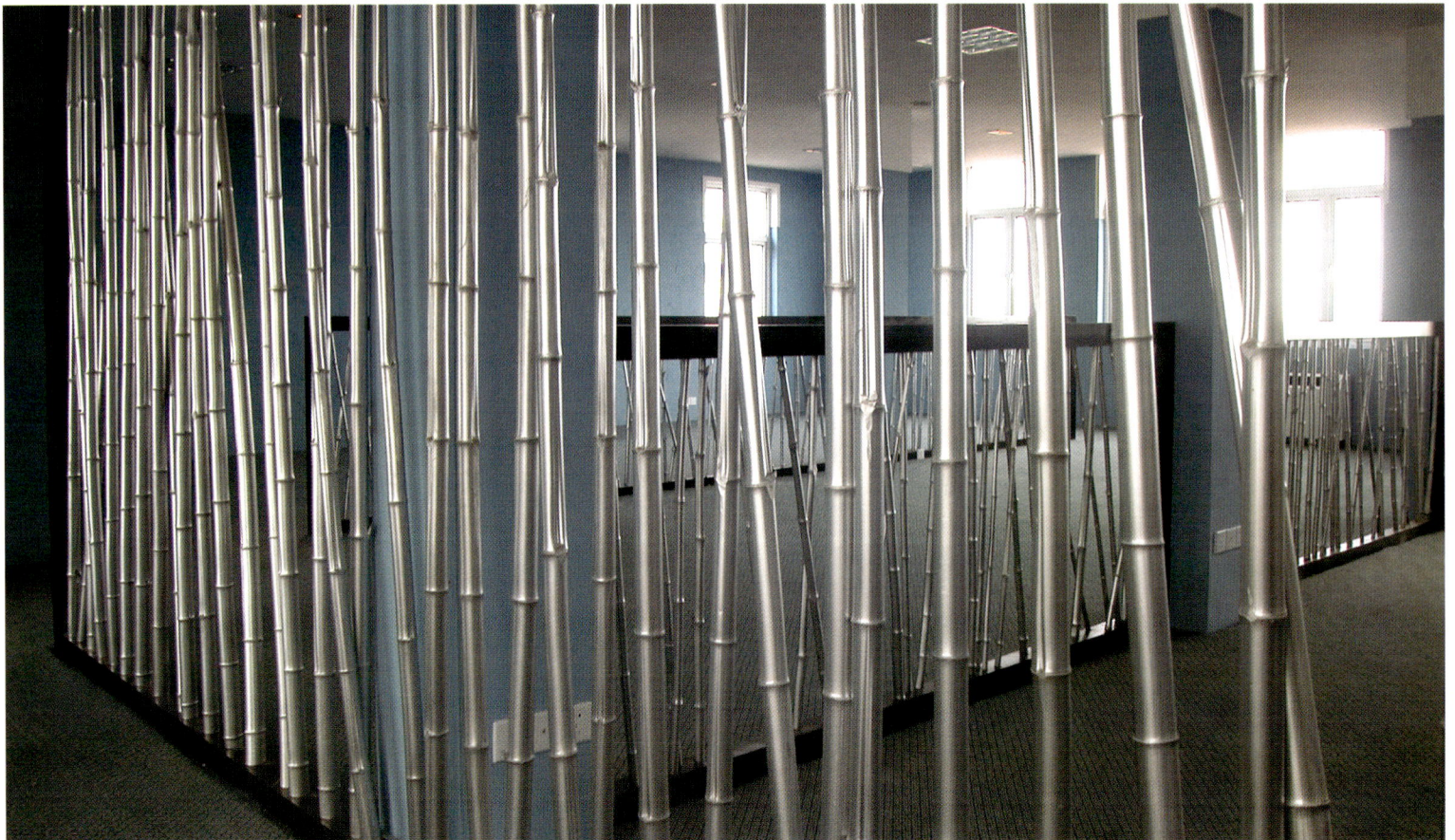

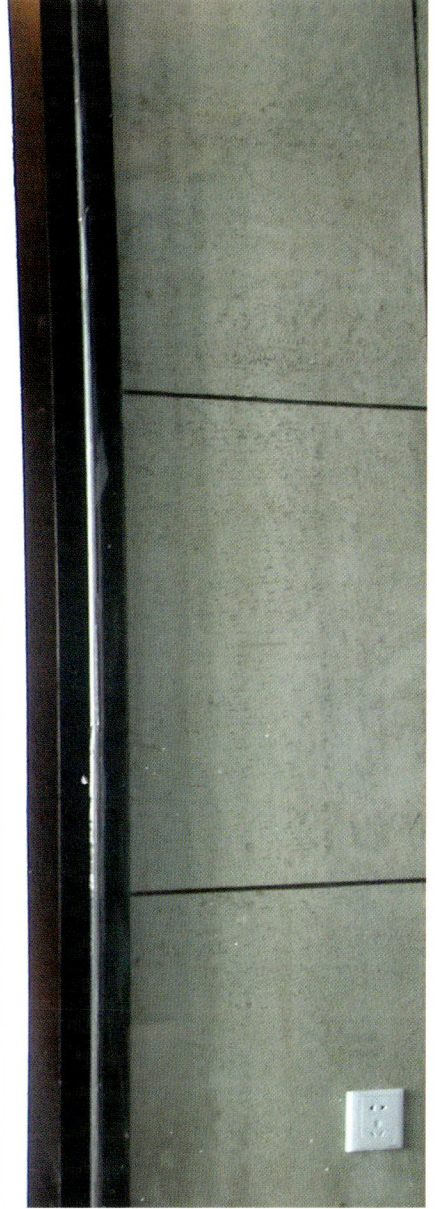

Cheongin

青仁

地　　点：香港
面　　积：500平方米
设 计 师：梁兆新
设计单位：新贤维思设计顾问有限公司

The coverage area of Hongkong Cheongin office is about 500 square meters. This design is combined with the design of product exhibition hall. The designer regards "protruding products" as the primary "design element", and combines the scattered products as a whole, integrating the enterprise's image into the products exhibition through coherent space arrangement.

In order to protrude the product, the designer combined the whole background with the white color, and chose black to design the chair. By adopting an atmosphere of white color, he contrasted brightness with white, hence didn't "disturb" the product's popular design color and pattern.

The designer used yellow to decorate the corridor, adding dignity to the atmosphere of the whole space. A modern colored product, a red ovate shape chair and attached tea table make guests feel the host's warm welcome.

The transverse white surface expresses its unique mystery and romance through the reflection of glass.

香港青仁企业办公厅，有500平方米的面积。这是一个结合产品展示厅的设计。设计师将特显产品作为首要"设计元素"，将零散的产品"化零为整"，并以连贯性的空间安排，将企业形象直接熏陶到产品展示中。

为了突显产品，设计师大胆地将整个背景统一到白色的系统中，把椅子设计成黑色，也是采用了无彩的中性色，但又巧妙地与白色只是形成明度的对比，从而使产品流行设计的色彩与图案没有受到"打扰"。

走廊等局部的黄色的运用，使整个空间的气氛平添几分贵重。一幅现代的色彩作品与一个红色卵形椅和连带的圆形小茶几，让来到的客商感到热烈的欢迎气氛。

横向断开的白色板面，通过玻璃而影射出特有的神秘与浪漫。

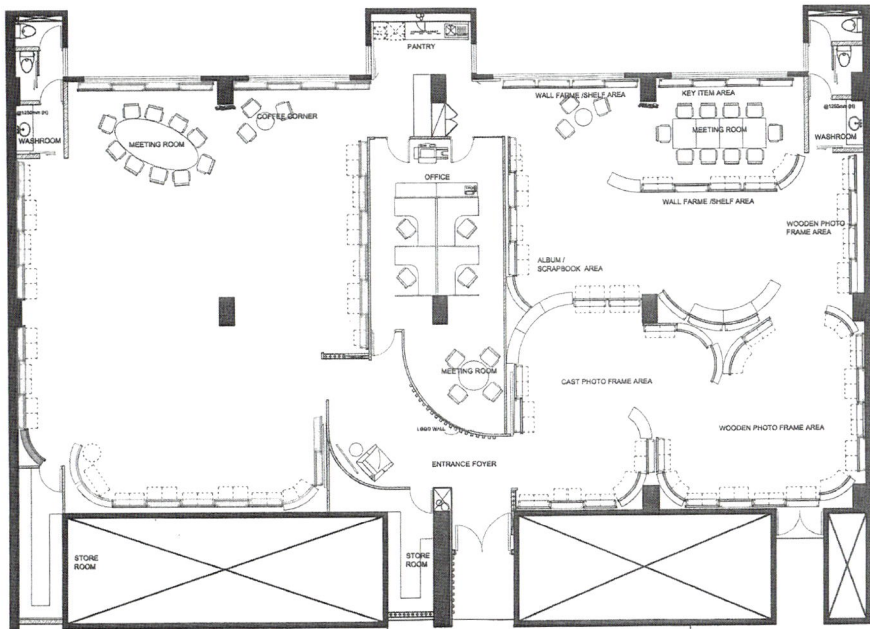

Floor plan labels:
- PANTRY
- WALL FARME /SHELF AREA
- KEY ITEM AREA
- WASHROOM
- MEETING ROOM
- COFFEE CORNER
- MEETING ROOM
- WASHROOM
- OFFICE
- WALL FARME /SHELF AREA
- WOODEN PHOTO FRAME AREA
- ALBUM / SCRAPBOOK AREA
- MEETING ROOM
- CAST PHOTO FRAME AREA
- WOODEN PHOTO FRAME AREA
- LOGO WALL
- ENTRANCE FOYER
- STORE ROOM
- STORE ROOM

Lzend Fashion Office Building

仟代服饰办公楼

地　　点：嵊州
面　　积：2500平方米
设 计 师：陈品浩
设计单位：宁波柏天设计公司

In this project, the designer made use of modern design methods to demonstrate how traditional Taoist aesthetic philosophy being used to decorate office space. The designer used "black" and "white" to be "yin" and "yang", emphasizing the abstract and the concrete, contrast of density and the variety of quantity. "White" can be "abstract"; "black" can be "concrete"; water flow can be river; deadwood of the tree can be forest... Black and white are alternating, varying and interpenetrating to make efforts to build a "Taoist" space which is constructed by the abstract and the concrete.

在本案的设计中，设计师用现代的设计手法来演绎传统道家审美哲学在办公空间的运用。

以"黑"与"白"互为"阴"和"阳"，强调虚实、疏密对比及量比的变化。"白"可以为"虚"，"黑"可以为"实"，一丘细水可以为江河，一树枯枝可以为山林……黑白交替，变化，穿插，着力营造富于虚实构成的"道"韵空间。

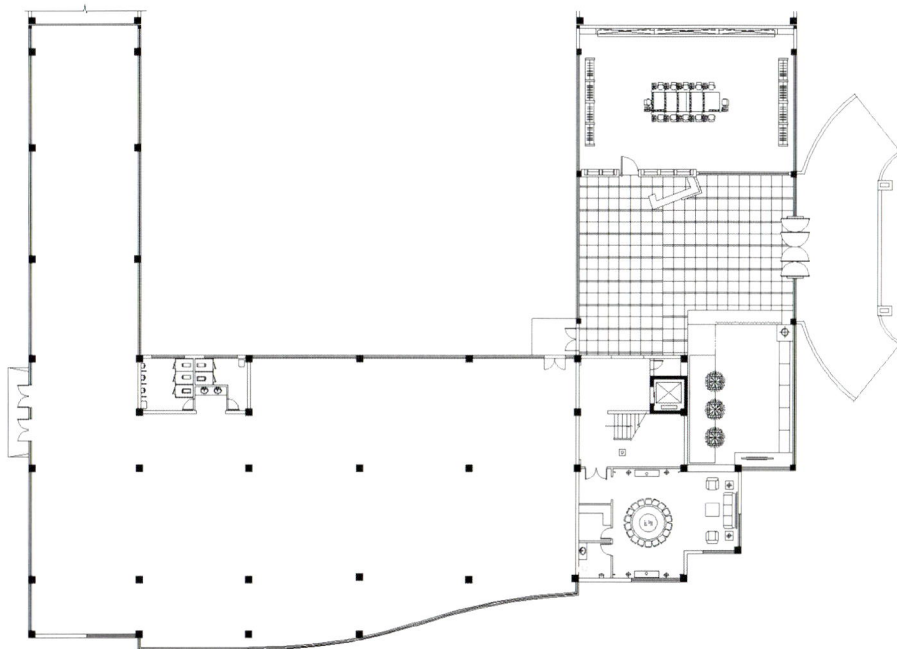

EURO

EURO

地　　点：香港
面　　积：960平方米
设 计 师：邹赞文、zenki lee等
设计单位：荟萃壹建工程有限公司

In a rigid cool space which is made of stone, tiles, and sand gray cement coating, designers proposed a red reception desk in the entrance, melting the rigid cool feeling with gentle and soft fire. Behind the reception desk, there is a huge luminescence light box of advertisement which brings hope and light to the customers. However, the mark of the company which is lighted by the light box, has outstanding colors but not too wildly known. The LCD Computer, the contacting phone and the red goose style soft sofa for resting-all of these could give the customer who comes for business the impression of warmth and high efficiency. A pot of graceful fresh flowers will add the romantic and warm atmosphere to the whole space.

Walking into the office, you will then see each room through the corridor. The white parted dragon bones and the big-piece glass installed through the partial crossband shaped sandblasting make the space apart from each other but connecte with each other. They are divided but transparent. The droplight on the ceiling of the corridor, decorated with the same distance arrangement, will give people an extremely strong rhythm sense. Combined with the glisten light made by the tile floor, the space becomes more interesting.

Designers adopt three neutral colors---white, gray and black---to decorate the whole space. The tables in the meeting room and the talking room adopt the natural color of wood, which would increase the feelings of nature, delight and affinity.

在一个由石材、瓷砖、水泥砂灰涂料界面围成的冰冷僵硬的空间中，设计师在入口处设置了一个红红的接待台，用柔情的火焰去融化这僵冷的感觉，并在接待台后立起巨型尺寸的发光广告灯箱，给客户带来希望与光明。而灯箱上映衬出的公司标志，颜色醒目，却不张扬。液晶电脑、联系电话、可休息片刻的鹅红式软沙发，都给来办事的客人留下热情与高效率的印象，而一盆雅致的鲜花则给整个空间增添了浪漫与温馨的气氛。

走进办公室，通过长廊进入各个房间。各空间的分割，通过白色的隔断龙骨与配装大片玻璃，通过局部横条状喷砂，使空间隔中带连，分而带透。尤其是长廊顶部的吊灯，通过等距的排列，给人极强的节奏感，再加上通过地面玻化砖的反光，更是相映成趣。

整个空间采用白、灰、黑中性色设计，只有会议室与洽谈室的桌子采用了木本色处理，以增加自然、愉快与亲和力。

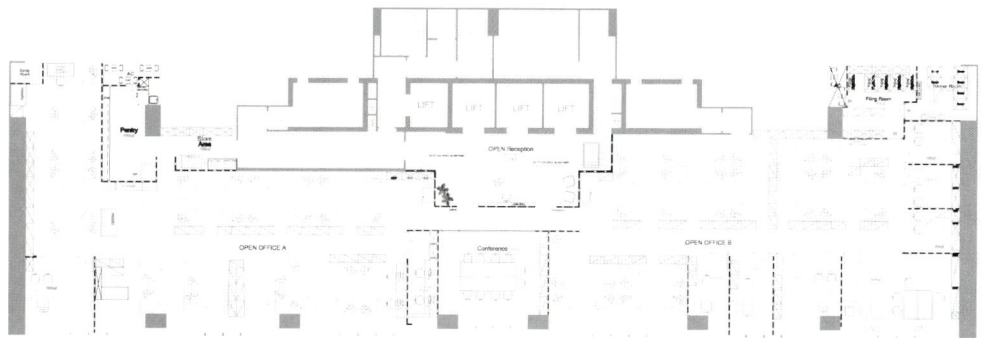

POAD

POAD

地　　点：香港
面　　积：1100平方米
设 计 师：邹赞文、zenki lee等
设计单位：荟萃壹建工程有限公司

When walking out of the elevator, the guests will see a wall whose length is more than 10 meters. In order to lead the guests, designers firstly transversely divide the wall into four parts, making the space become "short", and increase the kindness between each other. Secondly, they make a 10-meter-long light-emitting diode (LED) video-wall, which makes the vision of the space become capacious. The changing pictures and colors, combined with the company's mark, attract the guests who are walking along the long corridor to reach the reception desk.

Through setting the deep color carpet to decorate the corridor, the designers restricted the reception space. In addition, the ceiling's green shining coffering, the green reception desk, the company's green mark, echoing with the two green ribbons of the corridor, contrast and cooperate with the big scale's light yellow. In the office, khaki, chocolate, and grey green colored carpet, the beech colored central wall, the exuberant green plant and the deep colored office furniture not only harmonize with the former space's color, but also with the increasing steadiness, help people to finish the psychological preparation and transition successfully.

当客户走出电梯，正好面对的是一堵十几米长的墙，设计师为了引导客户，首先在高空间的墙身上进行了横向分割，分割成四段，使空间"变矮"，增加了彼此之间的亲切感。二是，在视线高度制作了10米长的LED发光二极管的电视墙，使空间的视线得以延伸。变换的画面与色彩，与公司的标志配合，吸引客户沿着长长的廊厅，通过浏览有趣的视频作为过渡，来到公司的接待台处。

接待处的空间，设计师通过在地面设置颜色深于廊厅花岗石的地毯，进行了限定。另外，顶棚的绿色发光藻井，绿色的接待台，绿色的公司标志，与长廊上下两条绿色的彩带遥相呼应，在淡黄色的大面积色彩的衬托下，在协调中进行着对比。

走进办公室，土黄色、赭石色、灰绿色相间的斑斓地毯，榉木本色的视觉中心墙板，郁郁葱葱的绿色植物，深色的办公家具，既与前置空间色彩协调，又有深一层次的稳健，成功地完成人心理上的铺垫与过渡。

POAD
SMARTER OUTDOOR

POAD GROUP LTD

POLARLINE DEVELOPMENT LTD

POLARLINE (CHINA) OUTDOOR ADV

POADESIGN COMPANY LTD

POADTAXI ADVERTISING LTD

POADmedia LTD

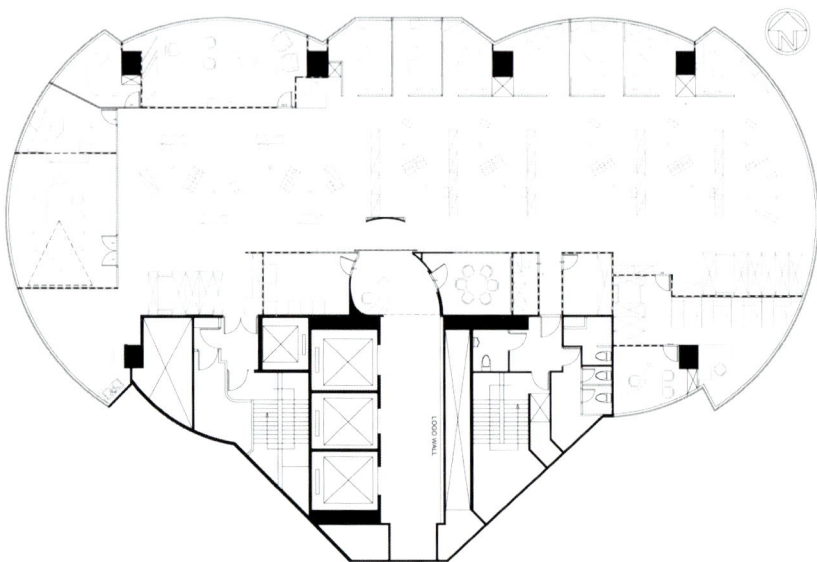

Beijing ViVi Audio & Video Decoration Co., Ltd.

北京畅想视听有限责任公司

地　　点：北京
面　　积：285平方米
设 计 师：张永雷
设计单位：北京东易日盛装饰有限责任公司

Beijing ViVi Audio & Video Decoration Co., Ltd. is the first company that undertakes the business of household video decoration in Beijing. At the request of the customer, this design is supposed to be modern and should present the space of household videos. Therefore, the designer applies the technique of combining the business space with household space, and the major difficulty lies in the knowledge of electronic and video products. In the design of the two audio-visual rooms including the wiring, the arrangement of lighting and the professional materials of audio-visual rooms, the designer made thorough communication with the professional designer of the audio-visual room for the purpose of the optimal audio-visual effect.

The materials chosen fully display the designing concept. On entering the room, a grand front desk is seen, and along with the black lacquered glass which creates a strong visual impact, presenting the fashionableness and capability of the company. The carpet on the floor adds more comfort to the space, and the warm-colored wallpaper with big flower patterns along with cream curtains connects with the designing concept of a cozy family, and adds a sense of hospitality to the space. It is very comfortable to enjoy the audio-visual effects in the bigger audio-visual room of the two, which fully combines the feeling of comfort and coziness. The smaller one, which applys a modern style, is delightful and interesting.

　　本案是北京市第一家承接家庭视听空间业务的公司。根据客户的要求，设计既要有现代感，又要体现家庭视听空间的感觉。因此，在做这套设计时，设计师运用了商业空间与家庭空间相结合的手法，但最大的难点是在电子产品与音响产品的知识上。在设计这两个视听室时，设计师充分与视听室的专业设计师沟通，使之达到最理想的视听效果，包括在施工中的布线、视听室光线的处理、视听室的专业材料。

　　在材质上，充分体现了设计思想。一进门，映入眼帘的是大气的前台和整个墙面的黑色烤漆玻璃，给人一种视觉冲击，体现公司的时尚化和公司的实力。地面的地毯增加了空间的舒适性，墙面暖色的大花纹壁纸和淡黄色窗帘将家庭温馨的设计理念结合进来，增加了空间的亲和力。两个视听室，大视听室完全是结合了温馨舒适的感觉，让人在这个空间里舒适地享受视听的效果，小视听室则结合了现代的手法，让人觉得趣味性更强。

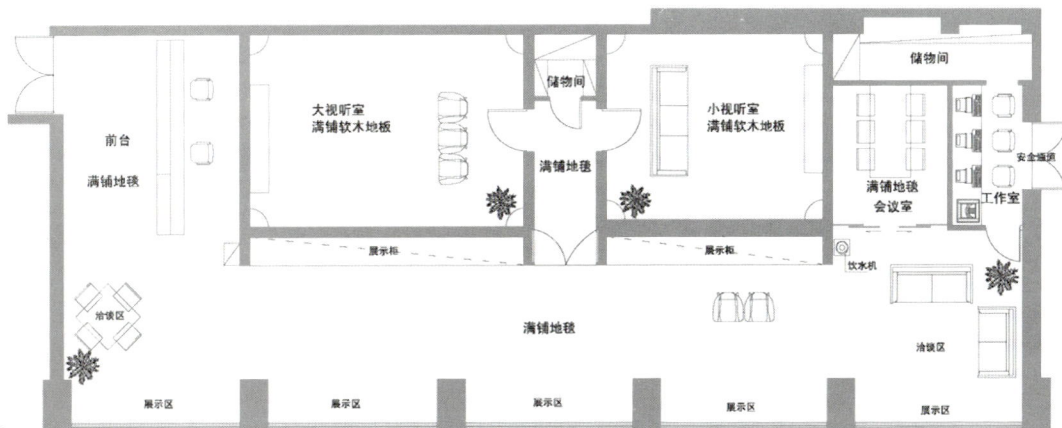

Shenzhen Linsn Technology Development Co., Ltd. Office

深圳灵星雨科技公司办公室

地　　点：深圳
面　　积：500平方米
设 计 师：陈颖
设计单位：深圳市秀城环境艺术设计有限公司

This company owns the world leading LED display core controlling technology. It has the common features of high technology companies in Zhujiang Delta, China: small scale, couple working, men will be responsible for technical research and development; women will be in charge of marketing and administration. But the controlling module they produced has taken up nearly 60% of the market of the same trades in the world.

By putting an interlayer in this 6.6m space, the two layers will be connected by a stair which contains the production, researching and developing, marketing departments etc. The design of the public area is put in a very important position because it is frequently used by visitors and developers who stay in the office for the longest time.

The meeting room on the first floor can be used as a canteen for staffs and café to drink coffee. The contrast feeling on visual sight of the washing room is very strong. The adding of a stone wall and crossband tables makes the space vivid and strong.

该公司拥有全球领先的LED显示核心控制技术，具备中国珠江三角洲地区高科技小企 业的共同特征：规模小，夫妻档，男的管技术研发，女的管市场及行政。但是其开发生产的控制模块已经占领了世界同类型产品市场的60%以上。

在一个6.6米净高的空间中加入一个夹层，楼梯连接上下两层空间，安排下生产、研发、市场等部门。公共区域的设计放在较重要的位置，因为那里是在办公室呆时间最长的开发人员以及来访问者经常利用到的地方。

一楼的会议室平时可作为公司员工的餐厅及品尝咖啡之用。公共洗手间区域显得视觉上的分量对比感觉比较强烈。因为有了一堵石头墙和横条虚格的加入，使空间显得生动而且有力。

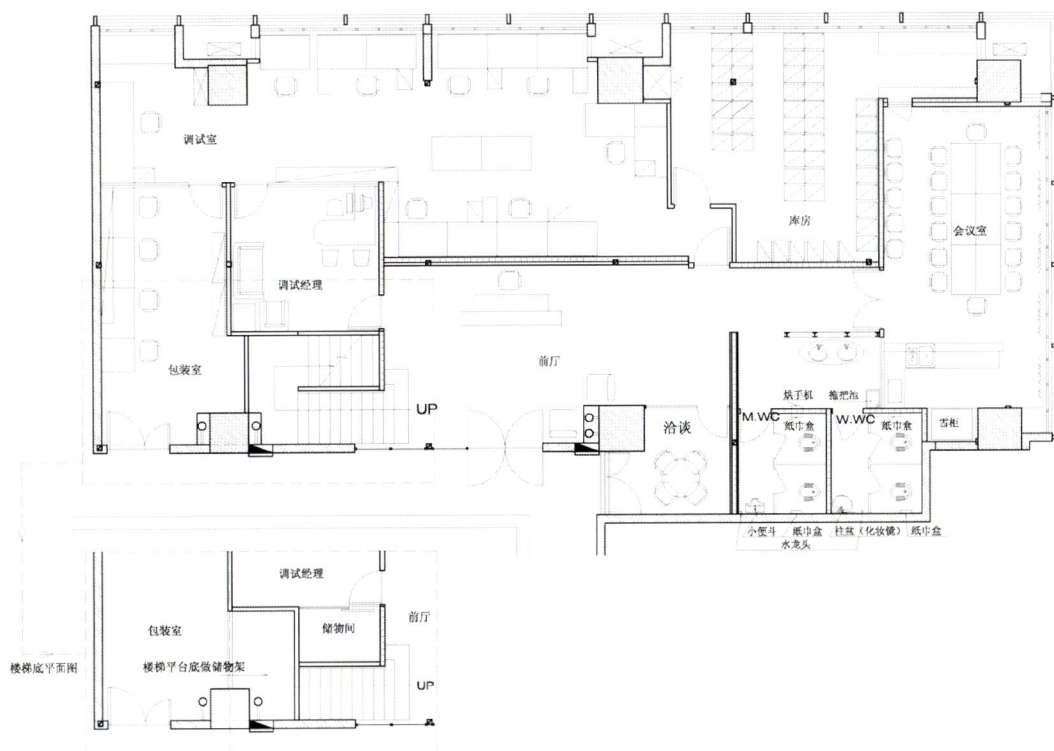

調試室

調試經理

包裝室

前厅

UP

库房

会议室

烘手机　擦把池

M.WC　纸巾盒　洽談　W.WC　纸巾盒　吊柜

小便斗　纸巾盒　柱筑(化妆镜)　纸巾盒
水龙头

調試經理

包裝室　储物间　前厅

UP

樓梯底平面图

樓梯平台底做储物架

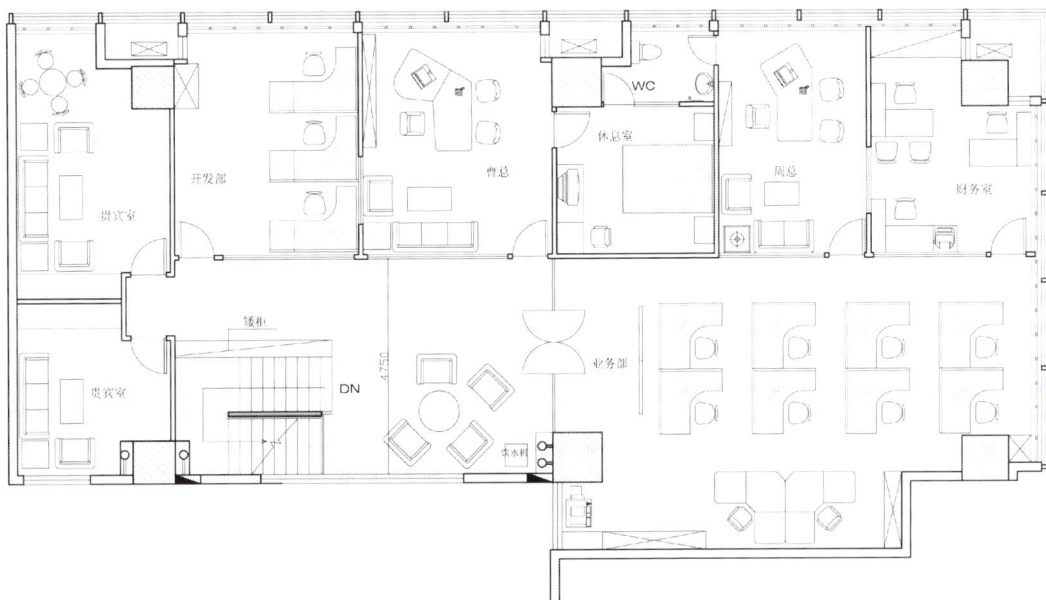

China US Venture Capital Group Co., Ltd. Office

中美风险投资集团有限公司办公室

地　　点：深圳
面　　积：900平方米
设 计 师：康华　杜伟
设计单位：深圳市汉筑装饰设计有限公司

In the spatial form, a character of briefness, sedateness and good taste is created, and the local cultural connotation is embodied properly. In the employment of materials, stones, vitreous tiles, aluminum boards, glass and metal components are the main materials. In the design of lighting, to provide a soft and pleasant lighting environment, most lightings are concealed and indirect illumination.

In the treatment of the interior space, a splitting arrangement of the construction is employed to form the partition and combination of environmental functions as well as the dynamic and quiet areas. In this way, the space is more open and easier to use, the stream lines are more reasonable, and the vision is more comfortable, so that a fresh, elegant, cozy, fashionable and pleasant environment is created. A harmonious unity of functions and visual effects is created through the combination of structures and materials, the selection and arrangement of colors, the decoration of details and the collocation of plants, and a new image of technological office is built, which is elegant, natural, vivid and modern. To meet both spatial and functional needs, compatible, comfortable and artistic furniture and ornaments are selected and arranged.

在空间形态上塑造简洁、大方、稳重的气质，并适当体现地域文化内涵。在材料运用上主要以石材、玻化砖、铝板、玻璃、金属构件为主。在照明设计上采用暗藏式间接照明较多，以形成柔和宜人的光照环境。

在内部空间处理上，利用结构的拆分布置，合理组织环境功能的分合与动静区域，使空间更开敞，使用更方便，流线更合理，视觉更舒适，从而形成了清新、典雅、舒爽、时尚的宜人环境空间。通过构造组合、材料组织、色泽选配、细节装饰和植物配置达到功能性与视觉效果的和谐统一，构筑了高雅、自然、生动、现代的科技办公的新形象。在家具、装饰品的选配上，根据空间与功能的要求，选用具有协调性、舒适性、艺术性的配置。

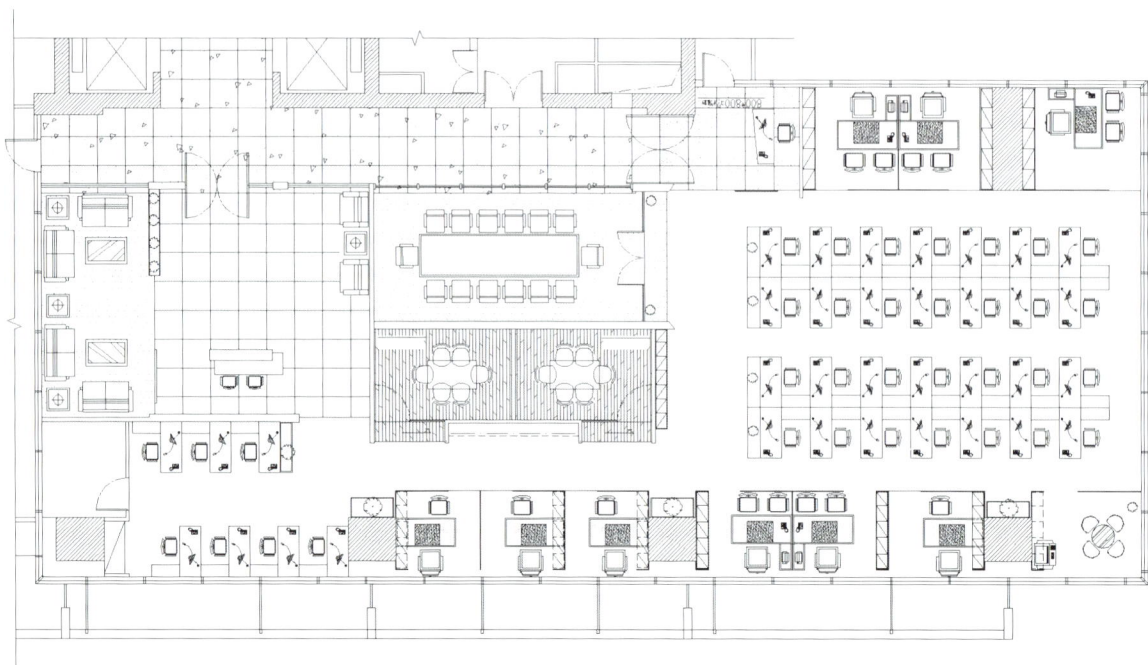

Inspiration Studio

汤物臣·肯文设计事务所

地　　点：广州
面　　积：470平方米
设 计 师：谢英凯
设计单位：广州市汤物臣·肯文设计事务所

"Focus on details and pursue breakthrough at any cost of untiring insistences and efforts." This is the motto of Inspiration Studio which was organized in the year of 2000.

In the designers' office, they have nailed the floor on the wall, or on the ceiling. They have refused to be confined within several skills or some kind of styles. They are actively searching for the potential element of the design object, making full use of its unique architectural order. They have been pursuing aesthetic sense from inner design to the outer, making the best of everything.

The capacious and bright designing space is convenient for the designers to communicate and cooperate. The long lines of fluorescent lamps provide enough lighting for the designers to draft the drawings. The designers want to create a "feeling area", which is pleasant to the eyes and very functional, and at the same time one can have the feeling of excitement and quiet.

The little potted green plant and the red swimming golden fish in the fish bowl, which are putting in the special resting area, bring the young designers inspiration and energy.

"追求细节、无限突破，从而为此付出不懈的坚持与努力。"这是成立于2000年的汤物臣·肯文设计事务所的座右铭。

在设计师自身的设计室中，把地板钉到墙面，钉到顶棚上，他们不想只局限于某几种手法或归属于某几种风格，而是积极寻找出设计对象潜在的因素，充分利用其独特的建筑秩序，在室内设计上力求由内而外地、充分地突出其特有的美感，并且达到物尽其用的目的。

宽敞明亮的设计工作空间，便于设计师们的交流与合作。长长的一排排日光灯管为绘制图纸提供了可靠的光照度。设计师要为工作创造一个"感觉的场所"，一个令人赏心悦目、功能齐备，同时又感到刺激和平静的环境。

特异的休闲地带，工作台面上的小小盆栽的绿色植物，鱼缸内来回游动的红红的金鱼，都给年轻的设计师们带来灵感与活力。

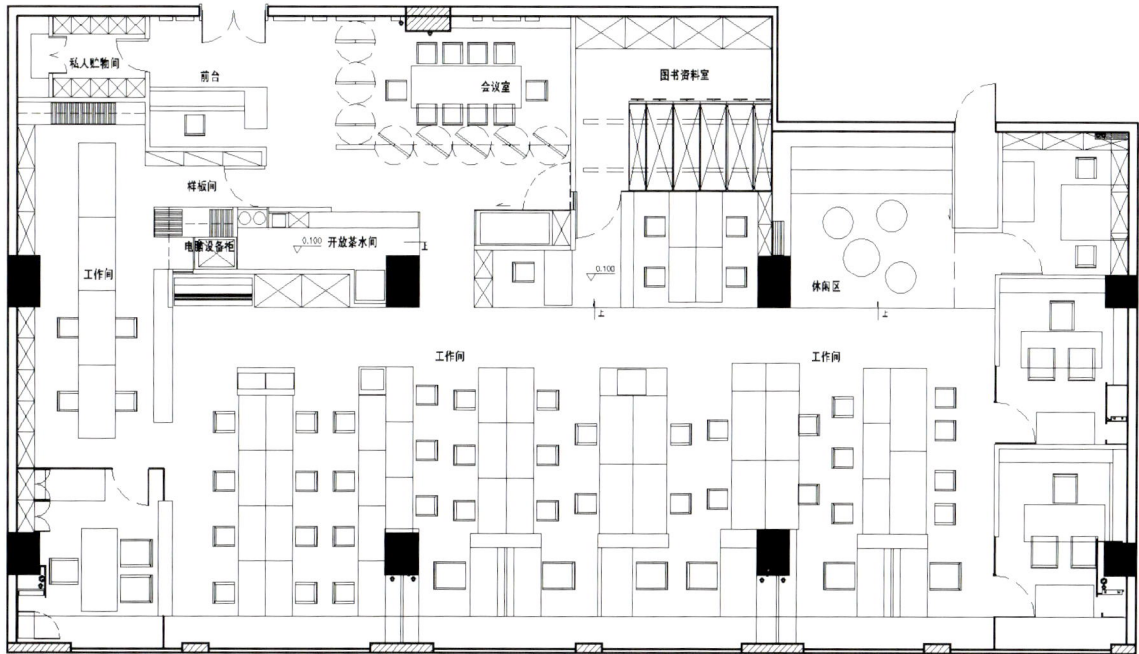

Yachang Visual Center

雅昌视觉中心

地　　点：深圳
面　　积：600平方米
设 计 师：熊华阳
设计单位：深圳市华旸环境艺术设计有限公司

This is a bardian space about 600m². The flexibility is the biggest feature of this design. The most obvious point of an advertisement company is that its work should be free and optional so that people can communicate conveniently and effectively. So the designer puts four moveable folding screens in the public working area which has been divided into four small working areas. Of course, you can divide it into many small areas by moving these four folding screens to satisfy the different requirements of the company. The folding screens are made of stainless steel and frosted glasses which are modern, light and handy. To go upstairs, the first thing you will see after entering into the company is not the usual front desk. There is only a wall left in this empty antehall with some letters made of log battens "IS ART". So we need not tell the feature of this company. The skew wall and log battens bring us a strong visual impact. Going round a rinsing cement folding screen, we have arrived at the open working hall. On both sides there are meeting rooms, studio and deliberating room. The rough rinsing cement wall, exquisite press polished steel and glass reflect with each other. And the frozen steel frame and warm log floor reflect mutually.

这是一个面积为600平方米左右的个性空间。空间的灵活性是本案的最大特点。广告公司的最大特点就是，工作需要自由随意、无拘无束，能够方便有效地沟通。为此，设计师在公共办公区域设计了四扇可滑动的屏风，它同时分隔出了四个小的工作区域，当然也可以利用屏风的自由组合来分隔各种不同的空间，满足公司不同的需要。屏风是由不锈钢和磨砂玻璃组成，现代而轻巧。拾级而上，一进公司大门，映入眼帘的并不是通常的前台。空空如也的前厅只剩下一面墙，上面有由原木条拼出来的"IS ART"几个字，公司特点不言而喻。斜斜的墙面和原木线条带来强烈的视觉冲击。从前厅绕过一堵小小的清水泥墙面屏风，就来到了开放式的工作大厅，两边则是会议室、摄影棚、研讨室。粗糙的清水泥墙面与精致的亚光不锈钢和玻璃交相辉映，冰冷的钢架与温暖的原木地板相互映衬。

China Mobile (Hong Kong) Researching Center Researching Office

中国移动(香港)研发中心研发部办公室

地　　点: 香港
面　　积: 2500平方米
设 计 师: 明光华　明罡
设计单位: 深圳市写意居装饰设计工程有限公司

The area of China Mobile (Hong Kong) Researching Center Researching Office is 2,500m². The designers use natural wood, concrete, steels, glass and dry wall to build a modern developing opening office room containing research, reformation, cooperation, training and discussion by using simple parallel and vertical lines. The steel glass partition takes the natural light into the office center. The whole space is very fluent and without any difference of place or limitation which shows the moving work. To keep the space of working, rest, meeting and training under the circumstance of free moving, it comforts the continual communication, adjustment, combination and the personnel change from offices all over the world inside the company.

　　中国移动(香港)研发中心研发部办公室占地面积为2500平方米，设计采用自然状态的木材、水泥、钢材、玻璃、清水墙，以简单的平行线和垂直线构筑了一个研究、革新、协作、培训、讨论的现代研发大开敞办公空间。

　　钢架高玻璃分隔墙将自然光线引入到办公室中心，整个空间流畅而全通透，无地域和界限之分，充分体现了移动办公。在个人可自由运行的情况下保留了办公、休息、会议、培训等空间，大大方便了公司内部频繁的交流、调整、组合以及与全国各地办事处人员的流动协调。

China UnionPay Shenzhen Branch

中国银联深圳分公司

地　　点：深圳
面　　积：1800平方米
设 计 师：康华　周辉
设计单位：深圳市华南装饰设计工程有限公司

The design of the office asks for relaxation but not formalization. The massive use of decorating wooden bars and marmoreal pointing joints, the height reduction of ceiling, the curving of glass, all together provide us an environment full of infectivity. There is a commodious reception center in the hall and small units for communicating. At the same time, direct or reflected light is used to create a better effect. As the working place uses central core tube as the main axis, the classical long and straight corridor will be the poison to this good environment. By the proper set of the public area and the working area, some delicate profiles are made out which can conceal the bald corridor. To achieve the effect expected, all the furniture and ordered scagliolas and metal sections are all designed elaborately. On the design of the gangway on the ninth floor, the designers use the combination of longitudinal wooden line and cambered ceiling to create a moving space which shows the originality of the designer. Even some ends of the walkways and the partitions between offices are designed to be very creative. Color, light and geometric figure connect with each other tightly and become a symphony of this whole place.

办公场所的设计力求放松、非形式化。大量地运用装饰木线和大理石的勾缝，降低天花，雕刻玻璃，共同产生出一种极具感染力的环境。既有较为宽敞的前厅接待区，也有洽谈的小单元。同时配合直接的或反射的灯光，以产生最佳的效果。由于办公场地以中心核心筒为主心轴，典型的长而直的走廊是良好环境气氛的毒药。通过对公共区域和办公单元的巧妙设置，产生出一些微妙的轮廓，掩饰了那种单调的走廊。为了达到预期效果，家具结构以及定做的仿云石、金属部件，都经过了精心的设计，在九层的过道设计上，设计师更是独具匠心，利用纵向的木线结合弧形的天花，创造了动感的空间。就连一些过道的端头，办公室之间的简单隔墙，都被设计得富于创造性。整个场地形成一个色彩、灯光、几何图形紧密衔接组成的交响乐！

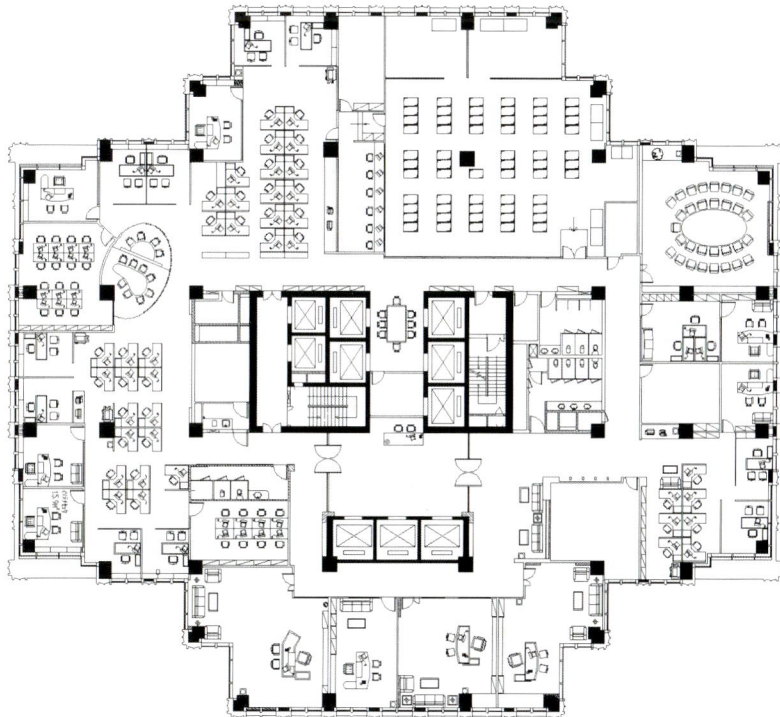

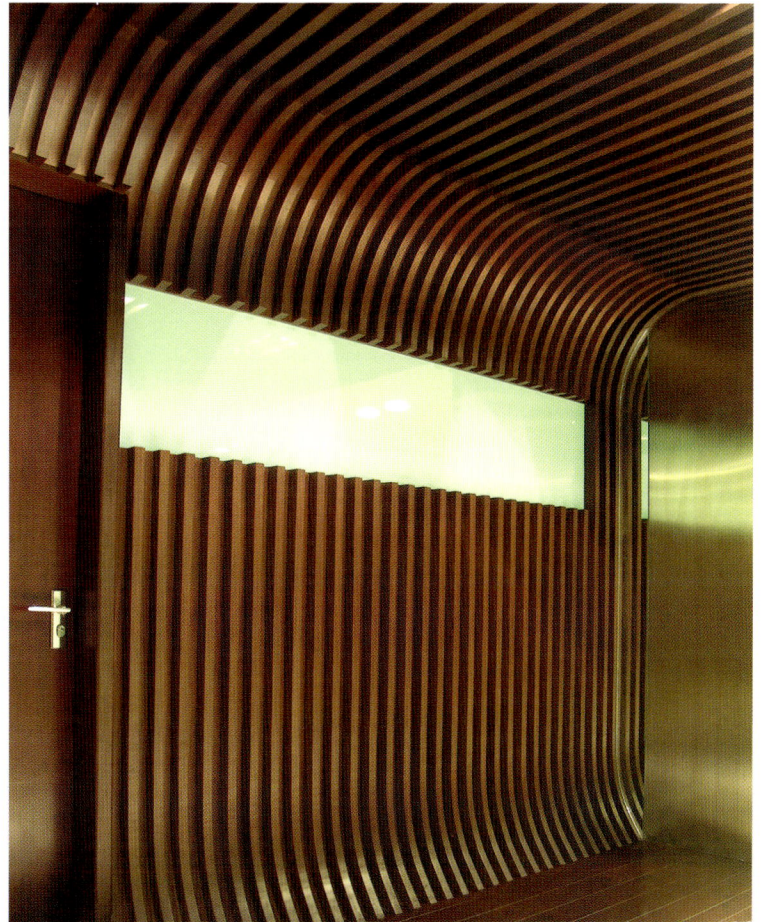

Tian An Digital City Innovation Science Technology Secondary East Building 905

天安数码城创新科技二期东座905

地　　点：深圳
设 计 师：狂狼阿呆
设计单位：狂狼阿呆室内设计厂

In order to represent the charm of modern high-tech office space, the designer adopts the electrical element welding "Printed Circuit Board", and sets undulated aluminium bunch line to decorate the part of ceiling. All of these settings adopt metallic grey to represent the charm of modern and science technology. The vertical design for the part of the wall and the use of red to decorate the wall and the part of the partition, bring the innovative team power of impulsion and attempt.

Under the slanting installed section steel glass partition, one dull landscape with white sand cobbles and water, several dry flowers in the pot, have increased several pastoral scenery for the inflexible space environment. What's more, in the transparent column-shaped resting area which is under the irradiation of the warm yellow light, the exhausted people can get drinks and food from the refrigerator. After simple treatments by the sideward machine, they may sit on the high chair, eating and chatting, and may have a short break.

　　为了体现现代高科技办公空间的魅力，设计师在墙面采用了电子元件焊接的"电路板"，顶棚局部设置了波纹铝金属束线管道铠装，这些装置都采用了金属灰的本色，彰显了现代与科技的魅力。局部墙板的竖向设计与墙面、办公隔断的局部红色的采用都给创新的团队带来冲动与跃跃欲试的力量。

　　斜置的型钢玻璃隔墙下面，一条没有干涸的白砂卵石枯山水，几束插在罐子里的干花，都为刻板的空间环境增添了几分田园景色，而在温馨的黄色灯光笼罩的空透的柱状休息场所，工作劳累的人们可以从冰箱中取出需要的饮料与食品，在旁边设有加工设备，进行简单的处理后，可坐在高高的吧凳上，边吃边聊，做一个短暂的休憩。

Ningbo Customs Office Stationed in Airport Secondary Integration Building

宁波海关驻机场办事处二期综合楼

地　　点: 宁波
面　　积: 2200平方米
设 计 师: 王寄明
设计单位: 宁波新世纪装饰设计有限公司

When looking at this exquisite modern building, people always remember the rhythm of the Eagles' "Hotel California", and remember their raucous singing. But the designer's original purpose is not to build a place where people could think of the past. He wants to build a place which is full of humanity's fashion, simple sports and leisure.

Actually, it is the moment when you come in that you can feel the spirit of modern age. In front the whole wall is a relief which made of stainless steel mirror. No matter the extending geometric surface which is made of the vigorous uneven shaped stainless steel, or the dazzling orange color which interpenetrated among them, all of them shocked people. The varying of stainless steel and its faint mirror effect bring people a visual shock of postmodernism. Compared with the common space of the first floor, the crush-room and guest room seem more private. Resting space for male and female adopts layered design style, making the space seems more private and convenient. The room for male adopts deeper warm color, concise outline, to show man's natural qualities of fortitude and deep. The room for female adopts quietly elegant cold color, with ornaments interspersed, adding the charming feeling, and revealing woman's character of gentleness.

望着这幢精致的现代建筑，不由使人想起老鹰乐队的那首"加州旅馆"的旋律和那沙哑的歌声。不过设计师的本意倒不是想把它打造成一个让人无法离去的怀旧场所，而是一个充满人文气息的时尚、简约运动与休闲之地。

事实上让人感受到现代气息是在进门的瞬间，迎面整堵墙面是用不锈钢镜面制成的浮雕，无论是凹凸不平、气势磅礴铺张开来的不锈钢几何平面，还是穿插其间的耀眼夺目的橘红色块，都给人扑面而来的震撼。而不锈钢的变形和略微模糊的镜像效果更给人一种后现代画派的视觉冲击。

与一楼的公共场所相比，楼上的休息室和客房更具私密性。男性与女性休息空间采用分层设计，使空间相对自由方便些。男性房间选用较深的暖色，线条简练，力展男性的刚毅深沉本色。而女性房间则采用淡雅的冷色基调，点缀上饰品，增添了妩媚感，烘托出女性温润如玉的性格。

Lifelong Educational Management Institute

终身教育管理学院

地　　点：深圳
面　　积：3200平方米
设 计 师：琚宾
设计单位：水平线设计

The speciality of this design is the thought of the arrangement of the function. It makes the design become many changeable directions. Each direction has a strong plane visual effect. The fix of natural and unvarnished material and the colorful and emulsified agglutinant form a fashionable but natural breath. The unimpediment design method and the facilities naked outside the ceiling make it a kind of lingering charm inside and the extensive sense of the space.

To pursue local contrast from harmony, the fashion of the furniture can reflect the century breath which can be consistent with the pursuit of enterprise and culture.

　　本案设计对平面设计规划的独特性是考虑到功能的安排。对功能的安排使其形成很多富有变化的角度，多个角度都会具有很强的平面视觉效果。材料上的自然质朴和与彩色乳化胶的搭配，时尚中透着自然的气息。无障碍的设计手法与天花裸露的设备使其具有建筑室内的韵味和空间的延伸感。

　　协调统一中追求局部的对比，家具的时尚性更能反映时代气息，与企业的文化追求相一致。

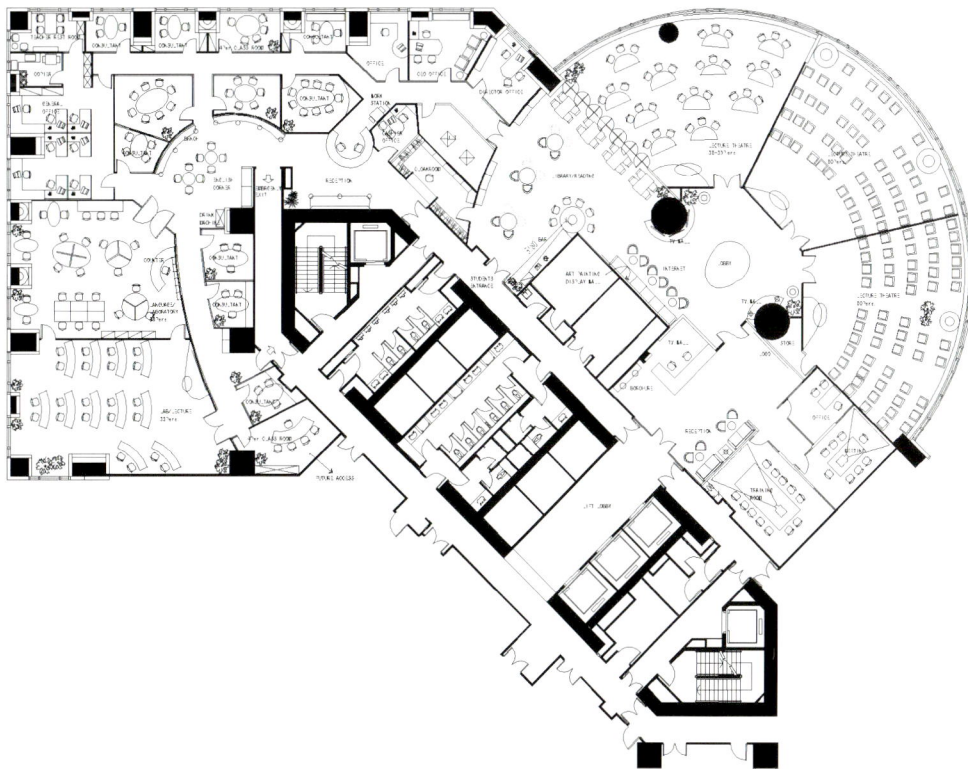

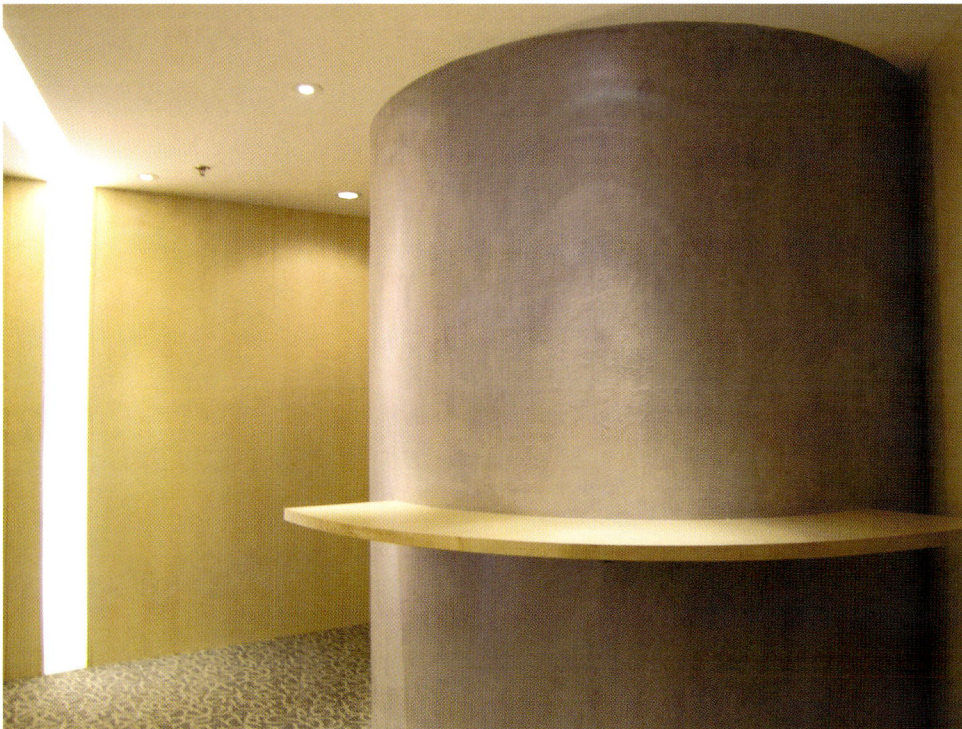

Ningbo Henghe International Trade Center Offices

宁波恒和国贸OFFICES

地　　点：宁波
面　　积：350平方米
设 计 师：宋国梁
设计单位：宁波红宝石装饰设计有限公司

This is a 350m² small working space. It was composed of some single small rooms and the profile was narrow and long. The square column densely arranged on both sides and some structural walls increased the difficulty of the layout of the space. To break the limitation brought by the narrow space, the designer uses very simple geometric figures such as square, circle and triangle to divide this space. These figures make each of the functional areas fluent and vivid. In the central area which is the cross point of the reception area，production exhibiting hall and public walkway, the round shape can make a natural transition and cushion to these three areas. The designer also uses the plasticity of the LG plastic to make three different colors to these different areas to divide and echo with local areas. For example, the triangle of the public corridor and the top have made a mutual reaction effects. In this design, the designer tries to put in more human cultural and geometrical elements. This breaks the single and straitlaced inertial thought for the space, and makes the design full of freshness, intelligence and human warmth.

这是一个350平方米的小型办公空间，原来由几个单独的小空间组成，呈"一"字形狭长，两边密排的方柱和一些结构墙给空间的排布增加了难度。为了打破狭长空间所带来的局限，设计师运用最为简练的方、圆、三角等几何图形对空间布局进行了分割划分，使各个功能区域顺畅生动起来。位于接待区、产品展示厅、公共走道交汇处的中心区域，圆的形体可以使三个区域得到更自然的过渡和缓冲。设计师还利用LG塑胶地材的可塑性，将地面根据不同的区域选配不同的色彩，以此来划分和呼应区域形体，如公共走廊的三角形就与顶部形成生动的互动效果。在本案中，设计师尝试给办公空间注入更多的人文因素和几何元素，冲破了办公空间单一、刻板的惯性思维，使该作品洋溢着清新、灵动和人性的温暖。

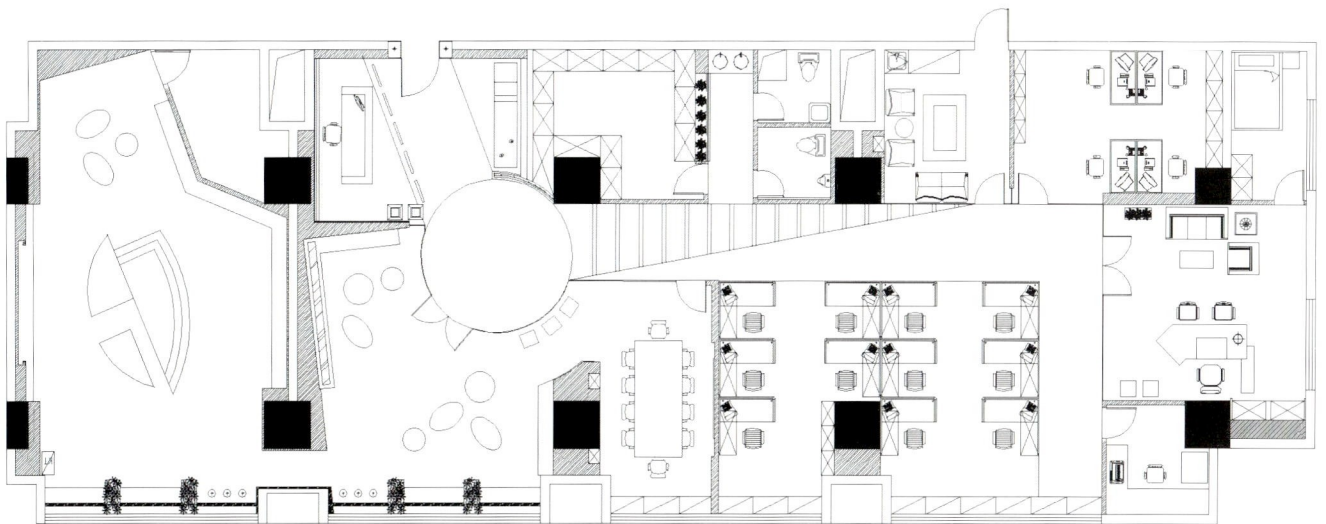

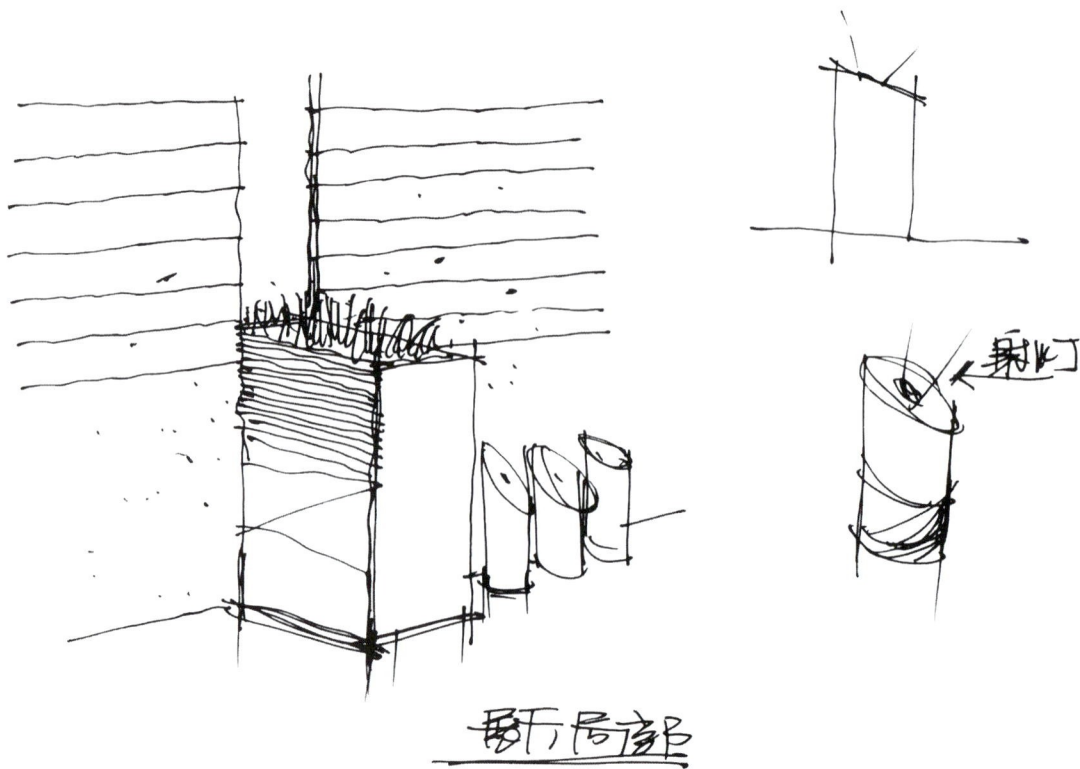

入口設計

絲草

排聚苯

花插

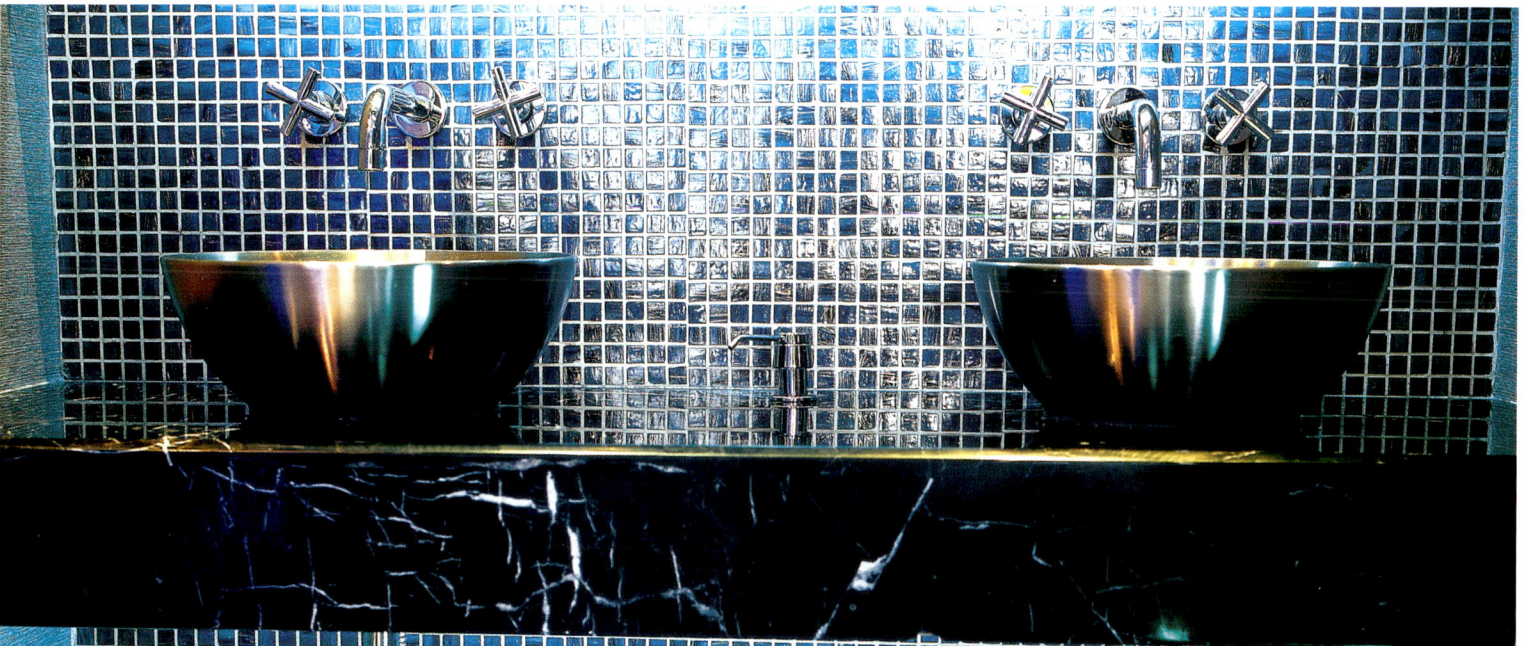